Between Theory and Observations

For other titles published in this series, go to
http://www.springer.com/series/4142

Sources and Studies
in the History of Mathematics and Physical Sciences

Steven A. Wepster

Between Theory and Observations

Tobias Mayer's Explorations of Lunar Motion, 1751–1755

 Springer

Steven A. Wepster
Department of Mathematics
Utrecht University
PO Box 80.010
3508 TA Utrecht
The Netherlands
s.a.wepster@gmail.com
s.a.wepster@uu.nl

Sources Editor:
Jed Z. Buchwald
California Institute of Technology
Division of the Humanities and Social Sciences
MC 101–40
Pasadena, CA 91125
USA

ISBN 978-1-4614-2503-8   e-ISBN 978-1-4419-1314-2
DOI 10.1007/978-1-4419-1314-2
Springer New York Dordrecht Heidelberg London

Mathematics Subject Classification (2000): 01A50, 01-02, 85-03, 37-03, 62-03

Printed on acid-free paper

Springer is part of Springer Science+Business Media (www.springer.com)

Preface

Several seemingly unrelated events in one's life may sometimes come together like the pieces of a jigsaw puzzle. Thus, a lifelong indoctrination with navigation, a part-time interest in astronomy and a late fascination with mathematics and its history brought me, after a seemingly aimless Odyssee, to a study of the so-called 'lunar distance method' of finding longitude at sea, and hence to the German polymath Tobias Mayer (1723–1762). His manuscripts, in particular those that witness his contributions to the practical implementation of the lunar distance method at sea, lay virtually uninvestigated in the State and University Library in Göttingen. Fortunately, the Mathematics Department of Utrecht University offered me the opportunity to literally figure out what Mayer's manuscripts contained, and to write my findings in a PhD thesis, defended in September 2007. The current book is an update of that thesis.

I want to express my thankfulness to all the colleagues, friends, and family who have guided and guarded me, in a very broad sense, on the voyage. I will mention only a few special people explicitly. These are my promoters Jan Hogendijk, who has an unrelenting trust in me, and Henk Bos, who teaches that historical sources and modern readers both deserve to be treated with the utmost esteem; also Liesbeth de Wreede, who is the best colleague that I could ever have wished for; and last but not least my father Arnout, for having been much more my pilot than I had imagined.

Figures 6.3, 6.4, 8.1, 8.3 and 8.4 have been reproduced with the kind permission of the Niedersächsische Staats- und Universitäts-Bibliothek, Göttingen. Figure 3.1 is courtesy of the Tobias Mayer Museum Verein, Marbach am Neckar. Figure 9.1 is reproduced from Mayer (1750a), courtesy of the Tobias Mayer Museum Verein, Marbach am Neckar.

Utrecht, The Netherlands
May 2009

Steven A. Wepster

Contents

List of Figures

List of Displays

Chapter 1
Introduction

1.1 Subject

Three great mathematicians dominate the history of lunar theory in the middle of the eighteenth century: Leonhard Euler, Alexis Clairaut, and Jean le Rond d'Alembert. Each of them made a lasting contribution to the theory of celestial mechanics and their results had a broader impact than on lunar theory alone. To name but a few examples, Euler codified the trigonometric functions and pioneered the method of variation of orbital constants; Clairaut solved the arduous problem of the motion of the lunar apogee, thereby dealing a decisive blow to the sceptics of Newton's law of gravitation; and d'Alembert worked out an accurate theory of precession and nutation.

But during the second half of the eighteenth century, the most accurate tables of lunar motion were those of Tobias Mayer; not because he was better at solving the differential equations of motion, but because he was eager to handle large and conflicting data sets in what could nowadays be called a statistical way. He was a true pioneer in the 'combination of observations', which comprises the handling of observational data in order to infer from them certain quantitative aspects of the physical reality. His tables of the moon's motion had an important application in the determination of geographical longitude at sea, by the so-called method of lunar distances.

What is known of Mayer's work on lunar motion? Alfred Gautier, the nineteenth century astronomer, included a brief analysis of the contents of Mayer's *Theoria Lunae* in his essay on the history of the three-body problem.[1] Delambre, although full of admiration for Mayer, dedicated only a few lines to it.[2] The twentieth century research by Eric Forbes, culminating in a biography of Mayer, has brought about something of a Mayer revival.[3] Forbes

---

[1] Gautier (1817, pp. 66–73).

[2] Delambre discusses Mayer's lunar theory in Delambre (1827, pp. 442–446), reviewed on p. 58 below.

[3] Forbes (1980).

S.A. Wepster, *Between Theory and Observations*, Sources and Studies in the History of Mathematics and Physical Sciences, DOI 10.1007/978-1-4419-1314-2_1,
© Springer Science+Business Media, LLC 2010

has constructed a unifying picture of the man Mayer and his work, after thorough investigation of correspondence, unpublished treatises and memoirs, and published sources as well. He produced an annotated edition of the correspondence between Tobias Mayer and Leonhard Euler[4] which provides intriguing glimpses of developing lunar theories, scientific life in the eighteenth century, and the relation between Euler and Mayer. Although Forbes' research is extremely valuable for anybody studying Mayer, he went astray in several aspects of Mayer's work on lunar motion: he left most of Mayer's manuscripts on lunar theory and tables uninvestigated and consequently his view on this important part of Mayer's work has to be corrected at some points.

On the other hand, a technique developed by Mayer to adjust a specific mathematical model to available observations is now reasonably well known among historians of statistics as one of the first successful attempts of nontrivial model fitting.[5] But the model is one of the lunar libration, not of its orbital motion, although it has been suggested that Mayer applied the same technique to improve his lunar tables. Whether this is true has never been investigated thus far.

Most other commentaries on Mayer's investigations of lunar motion are either based on the sources just mentioned or repetitions of frail anecdotes. Thus it becomes apparent that the past 250 years have seen hardly any research into the making of Mayer's lunar tables. In other words, until now we did not know how one of the most important lunar theories of the eighteenth century came into being. Harrison's work on timekeepers–the core object of the 'rival' method to determine longitude at sea–has attracted more historical research, I suppose partially because the principle of longitude determination using clocks is easier to explain than the lunar distance method, and partially because ticking clocks make more attractive museum exhibitions than almanac pages filled with numbers.

The goal of the present study is to investigate the evolution of Tobias Mayer's lunar theory and tables. A major point concerns the causes of the accuracy of these tables: on several occasions, Mayer had announced that their accuracy was a result of the adjustment of the theory to observations. However, he did not report his procedures in full, neither in publications nor in extant letters, and it remained unknown how he managed to fit a model involving tens of parameters to a corpus of over a hundred observations.

Before one undertakes to fit a model to observations, it is necessary to have a model in the first place. Which was Mayer's model? Was it the theory contained in his *Theoria Lunae juxta Systema Newtonianum*, which was published posthumously in 1767 by the Commissioners of Longitude in London? Most of the researchers in the history of celestial mechanics have skipped

---

[4] 31 letters have been preserved, and they have been translated into English, annotated, and printed thanks to the efforts of Kopelevich and Forbes, see Forbes (1971a).

[5] See e.g., Stigler (1986, pp. 17–24); further discussion of the technique will be found in Chap. 9.

Mayer's lunar theory on the assumption that it is a straightforward derivative of Euler's. However, their assumption is no longer tenable after Euler's and Mayer's theories are compared, as we will do in this study, although the former certainly had a strong influence on the latter.

The relation between Mayer's theory and tables is complicated and oblique. It was not, principally, his theory as displayed in *Theoria Lunae* that he adjusted to fit better to the observations. This has not been remarked before. An aim of our study will be to provide a new picture of the relations between theory and tables, as well as between the theory of Mayer and the theories of others.

Most of Mayer's manuscripts have been preserved after his death, including a number of treatises and memoirs most of which have been edited and published by now. The bulk of the manuscripts, however, consist of observations, calculations, (draft) tables, and similar materials unsuitable for publication. These papers, as dense with numbers as words are sparse, have hardly been studied, yet the manuscripts contain valuable information about Mayer's process of fitting. The lack of relevant information in published sources, and the abundance of unpublished manuscripts, clearly put the spotlight on the latter. The results of the present investigation are brought about mostly through detective work in Mayer's unpublished manuscripts. Though the work is not limited to statistics, it is in the spirit of what Stigler must have had in mind when he wrote:

> In studying the history of statistics, sometimes a question of conceptual understanding can be illuminated by a seemingly trivial computation. When writing in general terms, a scientist may be subject to a variety of interpretations, or seem to endorse a variety of methods of procedure. Yet a close numerical reworking of a statistical application by that same scientist may be revealing. A surprising twist may appear, or an unexpected limitation in scope of analysis may become apparent. Investigations into the exact numerical procedures a scientist followed can be frustratingly difficult, since the details are frequently absent, or errors of calculation or the typographer can confuse, leaving the historian to attempt speculative reconstruction. But when the exact results can be derived by a plausible procedure, by convincingly reconstructing in minute numerical detail what the scientist actually did, there can be an exhilarating sense of discovery and clearer understanding. For example, by reenacting Mayer's study of the geography of the moon, or Laplace's numerical investigation of the lunar tides of the atmosphere, or Quetelet's calculation of the propensities for crime, we gain insight into their conceptual understanding of statistics that could scarcely have been achieved otherwise.[6]

What emerges out of my study is a drastically changed view of the origins of Tobias Mayer's lunar tables and lunar theory, of their interrelationship, and of their relations to the lunar theories of Euler and–surprisingly–Newton. I also present new insights into Mayer's dealing with random observational errors and related statistical concepts.

---

[6] Stigler (1999, p. 80); see Stigler (1986) for the examples that he mentions.

1.2 Organization

The focus of this book is on lunar longitude; latitude and parallax are almost completely neglected. The perturbations of the latter two coordinates[7] are smaller, and less interesting phenomena are encountered in studying them, notwithstanding that they formed an important part of Mayer's work.

I have also kept out of this investigation several other aspects of Mayer's work, some of which concern lunar motion. To these belong some very interesting activities, such as an investigation of the variation of the eccentricity of the earth's orbit, his investigations of refraction and lunar parallax, research on the secular acceleration of the moon, and his critical analysis of some observations claimed by Ptolemy.[8] Inclusion of these aspects could serve to show with how much expertise and exertion Mayer went about the perfecting of his lunar tables, but it would distract from the main questions of their provenance and refinement. I have excluded virtually everything that has to do with instruments or observational practice for the same reason.

With the scope of attention thus limited, still plenty of interest remains. It is divided over the chapters in the following way.

Some, mostly standard, background to the why and how of lunar theory is provided in Chap. 2. A short biography of Tobias Mayer is sketched in Chap. 3. This biography is based on earlier work of Eric Forbes, but many new details of the development of his lunar research have been added.

Chapter 4 explains how to work with Mayer's lunar tables. Similar explanations are included with the published tables, but those are presumably not readily available to most readers. The chapter is included because some of the particulars of working with the tables play an important role in the sequel. The paramount characteristic of the tables is their multistepped nature, which I will explain in Chap. 4, but which is already defined in Sect. 3.4. It will be of concern most of all in Chaps. 6 and 7.

Mayer's lunar theory, written in 1755, is the subject of Chap. 5. I examine under what circumstances it was written, and I point out that its completion was delayed by difficulties which Mayer found hard to overcome. After a review of his theory, which has been wanting until now, I compare Mayer's theory to those of Euler (1753) and Clairaut (1752a). Although either Euler's lunar theory or his similar treatise on the so-called great inequality of Jupiter and Saturn were sources of inspiration, Mayer's theory is certainly not a slight deviation from Euler's, as has been supposed before. Clairaut's theory, on the other hand, most likely had a hitherto unrecognized impact on Mayer's.

---

[7] (Horizontal) parallax can be considered as a coordinate in place of distance, cf. Sect. 2.3.1.

[8] All of these subjects show up in the Euler–Mayer correspondence (Forbes, 1971a). The pointers to the manuscript sources are (see Sect. 1.3 for explanation of the references): eccentricity Cod. $\mu_{41}^\sharp$ pp. 51–52; refraction Cod. μ_{12} fol. 225–229, 236–238, Cod. $\mu_{30}^\sharp$; parallax Cod. $\mu_6^\sharp$, Cod. $\mu_{41}^\sharp$ fol. 14–17, 219–224, Cod. $\mu_{30}^\sharp$; secular acceleration Cod. $\mu_{41}^\sharp$ fol. 1, 2, 117, 155–178 (partly), 368; Ptolemy Cod. $\mu_{41}^\sharp$ fol. 145–146.

That theory of Mayer's leaves two things unexplained. The first of these is the multistepped format of his tables. Chapter 6 links, for the very first time, this peculiarity of Mayer's tables directly to Newton's lunar theory of 1702 (not to be confused with his *Principia*). I show that during 1752 Mayer developed the multistepped format from a reworking of the latter. This is a truly remarkable result because it implies that the role of mathematical analysis in Mayer's successful lunar tables is subtle and is in need of reconsideration.

When working out his own lunar theory three years later, Mayer had to justify this multistepped format, and he had to show that it agreed well with the theory. This may have been the most formidable obstacle responsible for the delay alluded to above. In Chap. 7, I reconstruct Mayer's way out. There I also disprove the benefit that Mayer claimed for the multistepped format.

Also unexplained in Mayer's theory is how he went about to adjust it to observations. He had always been clear that he did so, but most of his procedures have never been disclosed. Yet, the task was a daunting one, because there were more than twenty parameters and he used well over a hundred observations to adjust them. The computational tools for such a task were non-existent, also even a method or conceptual framework to approach it were wanting.

Mayer's manuscripts reveal that he developed a remarkable technique that is conceptually very close to our understanding of a *spreadsheet*. The quality of his tables depended mostly on his successful application of this spreadsheet technique to the process of fitting of the coefficients. Several aspects of working with these spreadsheets are studied in Chap. 8.

The topic of adjusting theory to observations brings up the subject of model fitting, and, related to that, of statistics. Astronomers were a precocious breed where this subject is concerned. Various aspects of Mayer's position in this field, which was but barely emerging in his time, are brought together in Chap. 9. Among these is a technique that several historians of science have proposed as Mayer's way to adjust the tables to observations, namely, the same one that he had exhibited in connection with his research into the libration of the moon. This is now known as the *method of averages*. I show that Mayer indeed tried to apply the very same method to the adjustment of tables to observations and that there are plausible reasons why he failed and abandoned the technique. It is even unlikely that Mayer saw it as a *method*.

The final chapter brings a revised view of the development of Mayer's lunar tables and theory, based on the results of the previous chapters. The most important new insights are highlighted, and there is also a summary of still open questions.

1.3 Conventions

We will frequently make use of a number of conventions, which will now
be explained. The technical terms that occur here are explained in Sect. 2.3.
Unless otherwise stated, I will follow Mayer and adhere to ecliptic coordinates,
using Mayer's notation for the four basic arguments:

ω longitude of moon minus longitude of sun
p lunar (mean) anomaly (except in Chap. 5)
ς solar mean anomaly
δ longitude of the moon from the ascending node of the lunar orbit

Depending on circumstances, these may relate to either mean or true motions.
Occasionally, we will need minor modifications of this scheme to accommodate the differences between mean and true motions, as will be made clear
in the text. As a fifth argument one might have expected lunar (mean) longitude, but we will have no need to assign a symbol to it. Since lunar true
longitude is the holy grail that the tables are supposed to yield, it will never
appear as an argument in those tables. We will also employ the following
astronomical symbols:

D the moon
$\odot$ the sun
$\oplus$ the earth
Ω the ascending node of the lunar orbit
Υ the vernal equinox

The word *equation* has acquired a meaning in astronomy quite different from,
and incompatible with, its usual mathematical significance. Apparently, the
words for 'correction' and 'equation' are similar in Arabic, and medieval trans-
lators of Arabic astronomical texts have accidentally used the latter where
the former was intended by the original writer. To differentiate between the
clashing meanings, I will write *equAtion* for the astronomical interpretation
as a periodic correction of the mean coordinates. The declashion will be
dropped in quotations. A similar distinction exists between mathematical
and astronomical *inequalities*. In its astronomical sense, an inequAlity denotes a deviation from uniform motion. InequAlity and equAtion are very
closely related concepts, even to the extent that they seem to be not always
properly distinguished, and the terms are sometimes mixed at random. The
former is associated with deviation from uniform motion, the latter with a
mathematical device to model such a deviation.

In the past, it was customary to measure anomaly of the planets and the
moon from the aphelion or apogee. Such was still the case in the eighteenth
century. Cometary orbits, however, once it had been recognized that they
formed closed ellipses governed by the same law of gravity, obstinately refused
to yield to the convention. Due to their elongated orbits, the tiny comets
were invisible along the outer part of their orbits; therefore, their anomaly

was naturally measured from perihelion. Laplace unified the planetary and cometary conventions in the beginning of the nineteenth century by taking the planetary and lunar anomalies with respect to the perihelion, respectively, the perigee. I see no reason to enforce the modern, i.e., Laplace's, convention onto the older astronomers, and I decided to follow their own convention. The difference in anomaly is 180°; therefore, in formulae expressing the equation of centre half of the signs in the expressions of the coefficients change.

This book is concerned with Mayer's lunar tables, their conception and development. Quite some detailed technical information will show up that is itself best presented in tables. To avoid obtuse and confusing language, I have decided that the word 'table' is reserved to refer to the historical specimens, while the ones appearing here in the text are referred to as 'displays'.

'Model fitting' refers to the process of adjusting the parameters in a mathematical model to observations. This is a modern notion which is referred to freely in this study. In no way do I intend to imply that Tobias Mayer thought of his activities in terms of 'model fitting' with its modern connotations. Rather, by the term 'model fitting' I will usually refer to the process of adjusting something that we now would call a mathematical model to observations. The mathematical model of lunar motion can take various guises: equations, sets of coefficients associated with those equations, or tables, i.e., equations in tabulated form. The process of fitting may always be assumed to apply to the parameters of the equations. Thus, when it is said that Mayer fitted (or adjusted) his tables to observations, it is implied that he adjusted the coefficients of the equations.

Mayer's work on lunar tables progressed quickly from about 1750 to 1755, and in this period several versions of his tables followed each other in quick succession. For ease of reference, I have assigned aliases to the many different versions of the tables, theories, and the like. I harvested the aliases from the rich vocabulary associated with the often fuzzy boundary of land and water in the Netherlands. Incidentally, the mariner who finds himself in waters where these names are applicable, is advised to guard not his longitude but his soundings. The aliases are set in this font. A complete list of all versions that I located in Mayer's manuscripts and printed matter is contained in Display A.1.

Mayer's manuscripts in Göttingen are referred to with symbols such as Cod. μ_1. Appendix C contains a list of these manuscripts with references to their classification in the Göttingen University Library. The same appendix lists the consulted manuscripts of the Royal Greenwich Observatory (RGO) archives now kept in Cambridge. References to the manuscripts normally follow the recto–verso convention: fol. 8r is the recto side of folio 8 and fol. 8v is its verso side.

Chapter 2
The Quest for Lunar Theory

Why was lunar theory so actively pursued? There were theoretical as well as practical reasons why astronomers, and especially those working in the middle of the eighteenth century, were highly interested in accurate methods to compute lunar positions for specified times. The theoretical ones had to do with the general progress of science in the eighteenth century (and later); on the practical side, lunar theory contributed to navigation at sea.

2.1 Scientific Significance

In his *Philosophiae Naturalis Principia Mathematica* (1687), Isaac Newton had laid down the general laws of motion of bodies, whether they are on earth or in the heavens. Newton introduced a new unified way of understanding motion, in particular that of celestial bodies.[1] He showed that the basic properties of the planetary elliptic orbits, which Kepler had found, were a consequence of the force of gravitation, which varies with the inverse square of the distance between attracting bodies. He recognized that the same gravitation was responsible for many other observed phenomena, such as the motion of comets, the precession of the earth's rotational axis, and the tides of the oceans. He also realized that the moon's orbit was affected not only by the gravitation of the earth around which it rotates, but also by the gravitation of the sun.

Although Newton could calculate several properties of the motion of the moon successfully from his laws of motion and the principle of gravitation, the moon seemed to deviate from its predicted path in some other aspects. For example, severe difficulties obstructed the theoretical derivation of the correct average advance of the line of apsides of the lunar orbit, i.e., the imaginary

---

[1] Currently, the most accessible edition is Newton et al. (1999).

S.A. Wepster, *Between Theory and Observations*, Sources and Studies in the History of Mathematics and Physical Sciences, DOI 10.1007/978-1-4419-1314-2_2,
© Springer Science+Business Media, LLC 2010

line that connects its apogee and perigee. To his dismay, Newton could get only half the observed value (about 40° per year) out of his calculations.

This discrepancy was serious. Newton's principle of gravitation was not yet readily accepted by everyone, partly because it implied the odd phenomenon of 'action at a distance' (masses exerting their gravitational forces instantly in every part of the universe, apparently without a material contact). Therefore, the principle of gravitation was put to the test: if it was really universally applicable, then it should explain all the motions of the heavenly bodies quantitatively. The problem of the apsidal line movement cast doubts on the theory.

In the 1740s, this problem was attacked by some of the ablest mathematicians of the time, including Leonhard Euler, Alexis Clairaut, and Jean le Rond d'Alembert. Their mathematical techniques were more powerful than Newton's, but nonetheless they, too, could account for only half the observed advance of the apsidal line. They proposed several causes for the discrepancy, including a law of gravitational attraction slightly differing from the inverse-square relation proposed by Newton. Eventually, though, Clairaut discovered in 1749 that some higher-order terms in the calculations, which he had supposed to be negligible, played havoc. With these terms taken into account, Clairaut computed a mean apsidal motion sufficiently close to the observed value. Clairaut's success contributed decisively to the general acceptance of the inverse-square law of gravitation, while his initial failure held a warning against a too easy neglect of supposedly very small terms.[2]

Outside of lunar theory, a similar difficulty was found in the movement of the planets Jupiter and Saturn, the two most massive planets in the solar system. From comparison of Tycho's observations and observations handed down by Ptolemy, Kepler had noted that the motions of these planets showed irregularities, the nature of which could not confidently be ascertained. Newton could calculate that these massive planets had a significant influence on each other's orbits, but it was beyond him to explain the phenomenon quantitatively. In 1748, this problem was made the subject of the annual contest of the *Académie des Sciences* of Paris. Euler's contribution[3] won the competition, yet he had failed to bring a satisfactory solution, and the Paris academy staged the same subject again for the contests of 1750 and 1752.

The importance of Euler's memoir lay in the new analytical techniques that he applied in it, including trigonometric series and an attempt to fit parameters to observations – we will encounter his memoir again on several occasions. Only in 1785 did Laplace show that the observed phenomena were indeed a consequence of gravitation. He found that the two planets are subject to perturbations of nearly 50′ and 20′, and a period of about 880 years, as a consequence of the nearly 5 : 2 ratio of their orbital periods. To discover this

---

[2] An extensive investigation of this episode is in Waff (1975); Clairaut published his results in Clairaut (1752a).

[3] Euler (1749a).

long-periodic effect, Laplace had developed tools to select potentially sizable terms out of the infinite number of terms that arise in the trigonometric series.[4]

A related topic, outside our scope but nonetheless worthwhile to be mentioned here, is the moon's secular acceleration. In 1693, Edmond Halley (1656–1742) compared lunar eclipses of recent, medieval Arabic, and classical Babylonian time and discovered that the moon's mean motion had been gradually increasing. Dunthorne in 1749 fixed this acceleration at $10''$ per squared century; Mayer first put it at $7''$ but later revised his value to $9''$.[5] The phenomenon defied explanation through the theory of gravitation, and many astronomers considered it as an effect of the aether: a thin hypothetical substance that was supposed to fill the universe.

It was Laplace again who showed in 1787 that the secular acceleration was actually a perturbation of very long period, caused by a periodical change in the eccentricity of the earth's orbit through the actions of the planets Jupiter and Venus. His success to account theoretically for the full observed effect lasted until John Couch Adams in 1854 discovered errors in Laplace's calculation, the correction of which caused half of the computed acceleration to disappear – and half of the observed amount to be unexplained again. It is now believed that the missing half is brought about by tidal friction, which causes the earth to spin down. The slowing of the earth's rotation increases the orbital velocity of the moon, and consequently the moon moves out to a larger orbit. Although Kepler's third law then implies a *decrease* of the moon's angular velocity instead of an increase, the moon's motion currently *appears* to accelerate relative to our slowing mean solar time.[6]

Adams's discovery of Laplace's error holds a warning to lunar theorists, comparable to Clairaut's discovery of the true precession of the line of apsides. The problems posed by the motion of the moon, and by the mutual influences of Jupiter and Saturn on each other, opened the research on the 'three-body problem', which has defied mathematicians in search of a closed-form solution except in a few special cases. Indeed, Bruns and Poincaré demonstrated that the general problem of three bodies does not allow the formulation of a sufficient number of relations between the coordinates and velocities of the bodies, in order for it to be solvable in closed form.[7]

---

[4] An excellent discussion of the history of this problem (and much more) is in Wilson (1985). As a side note, the simple ratio of the periods implies *that* sizable perturbations exist, but implies nothing about the relative magnitudes of the resulting perturbations.

[5] Mayer reported on his research in Mayer (1753b, pp. 388–390); his calculations are scattered throughout Cod. $\mu_{41}^{\sharp}$.

[6] Verbunt (2002). Incidentally, the same tidal friction mechanism has already slowed the lunar axial rotation to its state of stable equilibrium, synchronous with the lunar motion around the earth.

[7] Cf. Brown (1896, p. 27).

2.2 Application to Navigation

From a more pragmatic point of view, a good lunar theory was urgently needed in oceanic navigation and geography for the determination of geographical longitude. Whereas navigators as well as geographers had sufficient means to find their geographical latitude, sufficiently accurate methods of finding longitude were lacking. In fact, navigators had absolutely *no* way of knowing their longitude when out of sight of well-surveyed coasts (of which there were but few) except by keeping an account of courses and distances traversed. But with the instruments and procedures of the time, and the scant knowledge of ocean currents, the quality of this so-called *dead reckoning* was not so good that one would like to entrust a whole cargo's worth to it.

My own experience in small boat navigation tells me that an allowance for an error margin of 10% of recorded distance is not overly precautious—in twentieth century circumstances, that is. A seventeenth-century navigator would perhaps do well to allow for twice that error unless he was sure that there were no appreciable inaccuracies in his sea charts. Thus, to a party proceeding from Batavia to the Cape of Good Hope, which is a distance of some 5,000 nautical miles covered in perhaps two months, the suggested margin might have amounted to a week's sailing.

This systematic inability of the navigators of the sixteenth to eighteenth centuries to fix one of their coordinates on the earth became a concern of their kings and patrons. Substantial rewards were put in prospect, successively in Spain, the Netherlands, France, and England for solutions to the problem, neatly following the torch of sea power. Finding the longitude stood on a par with squaring the circle, according to many people. However, the longitude problem was overstressed: the attention that it received seems to be in imbalance compared with the lack of attention for some other potential causes of shipwreck.[8]

Probably the most well known of these rewards was stipulated by British Parliament as the *Act 12 Queen Anne, Cap. XV*, colloquially known as the *Longitude Act* of 1714. Their Act had certainly been inspired by the wrecking of Sir Clowdisley Shovell's fleet on the Scilly Islands and the consequent loss of the lives of 2,000 able-bodied seamen. This was one of the most dramatic disasters in British sea history; it was, however, caused by a navigational error in *latitude*, rather than *longitude*.[9]

The Act promised a reward of £20,000 for the discoverer of a method of finding longitude that was accurate to $\frac{1}{2}°$ of a great circle, i.e., to 30 nautical miles; lower rewards of £15,000 and £10,000 were to be granted for an

---

[8] Davids (1985, p. 86); and: '[...] in an age when the world was not yet properly charted the fact that one did not know the longitude of one's destination meant that the longitude of one's own position was less important than it would be today. [...] The dire results of a lack of means of observing the longitude have been overstressed by many writers.' (May 1973, pp. 28–29).

[9] May (1973, pp. 27–28).

accuracy of $\frac{2}{3}$° and 1°, respectively. The method had to be practicable and useful, both aspects to be demonstrated on a voyage to the West-Indies and back. A commission, which became known as the *Board of Longitude*, consisting of representatives of the Admiralty, Parliament, and scientists (including the Astronomer Royal and the Savilian and Lucasian professors of mathematics) was to oversee the execution of the Act. Many proposals were presented to members of the Board by various individuals, but the plans were almost always considered as impracticable, not useful, or even non-sensical, long before it was deemed necessary to bother the complete Board.

At least as influential as this British Act was the testament of the French politician Rouillé de Meslay of 1715, stipulating that the *Académie des Sciences* in Paris should organize an annual prize contest. The subject of the contest would alternately address the motions of celestial objects and navigation (in particular longitude finding). These contests formed a significant impulse for the work of Euler, Clairaut, and others who contributed their essays in response to the prize questions. Between 1720 and 1792, the lunar orbit and other, related, problems in celestial mechanics repeatedly provided subjects for the challenge.

In the course of about two and a half centuries (from the beginning of the sixteenth to the middle of the eighteenth), the four most seriously pursued candidate solutions of the longitude problem employed either *timekeepers*, *lunar distances*, the *satellites of Jupiter*, or the *variation of the compass*. The first three of these aimed at providing the local time on a (possibly distant) reference meridian: the timekeeper method by setting a reliable watch to that time, the lunar distance method by the relatively swift motion of the moon (to be further explained below), and Jupiter's satellites by their predictable crossings of Jupiter's visible disc or its shadow cone. The fourth method depended on the usefulness of a relation between magnetic variation and geographic longitude, but the relation is rather weak, even non-existent in some regions of the globe, and moreover slowly changing with time. Cartographers could – contrary to navigators – afford to wait for particularly favourable celestial events, such as lunar eclipses or occultations of stars by the moon, to determine longitude. I will now briefly explain the principle of lunar distances, in order to clarify the relation between longitude determination and Mayer's work on lunar motion.[10]

The method of lunar distances exploits the relatively swift movement of the moon, which completes its orbit in about 29.5 days with respect to the sun or in 27.3 days with respect to the stars. In other words, the moon moves roughly its own diameter (30′) per hour relative to the stars. Therefore, if

---

[10] For further reading on the history of longitude and lunar distances, see Andrewes (1996), Howse (1980), Cotter (1968), and Marguet (1931). The best manuals on the lunar distance method, in true nineteenth-century style, are contained in Graff (1914), Jordan (1885), Chauvenet (1863), Brouwer (1864), and Bohnenberger (1795). A refreshing modern approach is in Stark (1995), and an increasing number of modern-day enthusiasts can be reached over the Internet.

an observer can make an accurate measurement of the place of the moon relative to another celestial body (whose position is accurately known) and if the observer has accurate knowledge of the movement of the moon with time, then he can combine the two and derive the time of his observation. Metaphorically, it is as if the moon is the hand of a celestial clock, with the stars and sun forming the dial. The somewhat puzzling expression of 'taking a lunar *distance*' should therefore be understood in the sense of measuring the angular separation of the moon from another celestial body.

Quite a number of difficulties make the lunar distance method awkward to apply. Many of these are directly or indirectly connected with the inherent inaccuracy of the method: the moon revolves approximately 30 times as slowly around the earth as the earth about its own axis. Therefore, errors in the obtained lunar distance (i.e., after the measurement and after all necessary computations have been made) get magnified by the same factor of 30 in the computed longitude. Thus, if the lunar distance is off by $2'$, the longitude will have an error of $1°$, the lowest accuracy limit set by the Longitude Act. The contemporary enthusiastic reactions to Mayer's lunar tables were due to the fact that they were the first to predict lunar positions to within a $2'$ error margin; in other words, they were the first tables that did not consume already more than the allowable error margin even before the observer had taken his measurement.

The principles of the lunar distance and timekeeper methods had been published by, respectively, Johann Werner in 1514 and by Gemma Frisius in 1530. More than two centuries later *both* methods were brought to fruition by, respectively, the lunar tables of Tobias Mayer (presented to the Board of Longitude in 1755) and the timekeepers of John Harrison (of which the fourth and last was presented to the same Board in 1760).

I have often pondered the near-coincidental perception and the near-coincidental realization of these methods, separated by such a long time-span of about 240 years. The difficulty with the moon was its seemingly irregular motion, which had to be modelled mathematically; timekeepers too had their irregularity but theirs was to be harnessed by mechanical innovations. Werner and Frisius had conceived their ideas while Europe was expanding its horizon over the edges of the oceans, yet it took a dozen generations to expand the necessary technological abilities.

As a side remark, a particular role on the longitude stage was reserved for Edmond Halley. Halley was most instrumental in encouraging Newton to write *Principia Mathematica*, and subsequently in publishing that book, because he recognized the invaluable assets – if not foundations – that it was to offer to astronomy. But Halley had also spent several years as a sea captain: he had been surveying the geomagnetic field in the Atlantic Ocean (with an eye on its supposed potential for longitude finding), the coasts and tides of the English Channel, and the treacherous sandbanks of the Thames Estuary. When in 1675 the Royal Observatory was founded in Greenwich with the explicit object of improving astronomy for the sake of navigation

Fig. 2.1 Horizontal
coordinates on the
celestial sphere

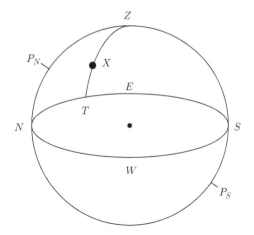

and in particular longitude finding, the nineteen-year-old Halley was involved.
Decades later, in 1720, he succeeded Flamsteed there as the Astronomer
Royal. Spending almost the final quarter of his long life in that position, he
must have had an extraordinarily diversified view of both the theoretical and
practical aspects of longitude finding.

2.3 Astronomical Prerequisites

The affiliation between mathematics and astronomy is long and intense, but
nowadays not every mathematician is versed in astronomical terminology.
This section is intended as a reference to those who feel uneasy (or perhaps
even lost) amidst the ecliptic, nodes, anomalies, and inclinations. It will guide
them through a little background in coordinates, elliptic orbits, and equations
of the lunar orbit.

2.3.1 Celestial Coordinates

A terrestrial observer looking at the night sky, perceives the stars and the
planets as if they are attached to a giant sphere of which he is the centre:
distances cannot be discerned without advanced techniques, only directions
can. This metaphor of the celestial sphere is an extremely powerful one and
will guide us in this section.

Figure 2.1 shows the celestial sphere with the comparatively extremely
small earth in its centre. The partially drawn line $P_N P_S$ is the earth's polar
axis extended, around which the sphere, with celestial body X attached to

Fig. 2.2 Ecliptic
coordinates on the
celestial sphere

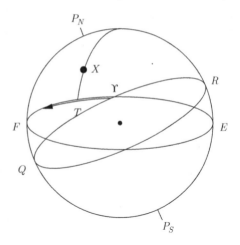

Fig. 2.2 Ecliptic
coordinates on the
celestial sphere

it, appears to rotate. Z is the zenith directly overhead of the observer and
$NESW$ is the horizon; these do not take part in the apparent rotation of the
sphere. The meridian circle $ZP_N N P_S S$ intersects the horizon in N and S,
which define the observer's directions north and south. The great circle arc
ZX extended intersects the horizon perpendicularly in T; the altitude of X
is the arc XT, and the azimuth of X is the arc NT, which must be labelled
E or W to indicate if it is measured in easterly or westerly direction. The
azimuth is not defined if X coincides with Z.

Thus, the immediately observable coordinates of a celestial body are its
altitude and azimuth, which are related to the observer's horizon and meridian. The coordinates of each body change constantly (except when coincident
with one of the poles) and it seems as if the celestial sphere rotates around a
fixed axis. We will make frequent use of this point of view, although we know
that in reality it is the earth that rotates.

Clearly, the celestial sphere must be provided with a coordinate system
independent of the rotation of the earth and of the location of the terrestrial
observer. As such, we will be concerned mostly with the geocentric ecliptic
coordinate system. It is geocentric, meaning that the centre of the earth is
taken as its origin. From this point of view, the earth stands still while the
sphere rotates daily around it.

Taking the earth at rest in the centre of the celestial sphere, we distinguish
between two different apparent motions of the sun. It moves with the apparent
daily rotation of the celestial sphere, which is accountable for its rising and
setting; besides, it appears to orbit around the earth once a year in the
opposite direction. The plane in which the annual motion takes place is called
the plane of the ecliptic. The ecliptic is defined as the great circle formed by
the intersection of this plane with the celestial sphere.

Similar to the ecliptic, we can think of the plane that contains the terrestrial equator, and intersect it with the celestial sphere to obtain the celestial

equator, or just equator for short. It is a great circle just as the ecliptic. These two great circles are tilted with respect to one another, forming an angle of about 23.5°, called the obliquity of the ecliptic. Twice a year the sun crosses the celestial equator: on or about 21 March, it crosses to the northern hemisphere (which is located above its terrestrial namesake), and it returns to the southern hemisphere about 22 September. At these instances, the sun occupies the intersection points of the ecliptic and the equator; these are termed the vernal and autumnal equinox, respectively, referring to the fact that on those dates the lengths of day and night are equal everywhere on earth.

Now we come to define the ecliptic coordinates. In the ecliptic, the vernal equinox provides a point of reference, commonly marked by the symbol Υ. Consider (Fig. 2.2) a celestial object X on the sphere (not one of the ecliptic poles). Let T be the point on the ecliptic closest to X, hence $XT\Upsilon$ forms a right angle. The coordinates of X are then its ecliptic longitude ΥT and its ecliptic latitude TX. The longitude is measured from Υ in the direction of the annual motion of the sun from 0° to 360°, as indicated by the arrow in the figure. Traditionally, it was customary to divide the ecliptic in twelve so-called signs of 30° each. A longitude of, say, $9^s1°$ equals 271°, and $3^s22°30'$ equals 112.5°. Latitude is measured from 0° on the ecliptic to 90°, north or south, the hemispheres being labelled according to the terrestrial poles.

The physics of gravitation oblige us to recognize that this mathematical coordinate system is in fact time dependent. First and foremost, the rotating earth is not absolutely spherical. It is more like an oblate spheroid, with excess mass at its bulging equator. Solar and lunar gravitation exert a net moment on this bulge trying to pull it into the plane of the ecliptic. Since the earth behaves as a spinning top, this moment makes it precess, i.e., the rotational axis rotates itself slowly around a mean position perpendicular to the ecliptic. As a result, the vernal equinox is not a fixed point, but it precesses backwards along the ecliptic at the rate of approximately 50″ per year relative to fixed space. Moreover, the moon's orbit is inclined to the ecliptic, and the lunar gravitational pull on the bulging terrestrial equator causes an additional wobble known as nutation, affecting the position of the vernal equinox periodically with period 18.6 year and amplitude 17″. Precession was discovered by Hipparchus in the second century BC; James Bradley detected nutation in 1728 but he announced it only in 1748, when he had verified its cause through a full period by observations.[11]

We have thus far disregarded distances. That is perfectly reasonable for our current purposes as long as only stars are concerned, because even from across two diametrically opposite positions in the earth's orbit (temporarily taking up the heliocentric view), shifts in stellar positions are at most 0.3″.[12]

---

[11] Bradley's announcement will be found in Bradley (1748).

[12] Bradley had been searching in vain for this stellar parallax and discovered aberration and nutation instead (aberration is an apparent deflection of light rays, resulting from the finite speeds of light and of the earth, which we will have no need to consider). Stellar parallax was first detected in 1838 by Bessel.

Fig. 2.3 Parallax

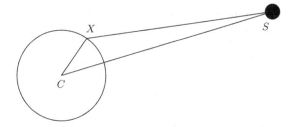

But for bodies in the solar system such as the sun and moon, we need to take their distances into account, in particular in relation to the size of the earth. This brings us to the subject of parallax.

Consider (Fig. 2.3) an observer X standing on the earth with centre C. The direction of the celestial object S as observed by X differs from the direction as it would have been observed from C, the origin of our coordinate system, unless S happens to be exactly overhead in the observer's zenith. The difference is called the parallax and it depends on the distance SC of the body, the radius CX of the earth, and the observed altitude of the body, $\angle CXS - 90°$. It is largest (ceteris paribus) when CX is perpendicular to SX, i.e., when the body is in the horizon of the observer; in that case, it is termed the horizontal parallax. The oblateness of the earth makes it necessary to define the equatorial horizontal parallax, which is the horizontal parallax for an observer located on the equator; the horizontal parallax at the latitude of X is then computed assuming that the earth is an ellipsoid.

With the observer thus ruled out, and the radius of the earth considered constant, the horizontal parallax P depends only on the distance of the object from the earth's centre, as expressed by the relation $\sin P = CX/CS$. It can now be seen that parallax can be considered as a coordinate instead of radial distance. There is also another way to look at horizontal parallax: to wit, that P denotes the apparent radius of the earth as seen from S.

The lunar parallax, i.e., the horizontal parallax of the moon, is about $57'$, the solar parallax is a little bit less than $9''$. These parallaxes should *not* be compared with the (annual) stellar parallax of $0.3''$ mentioned earlier, without considering that the base line in the former is the earth's radius, in the latter, however, the radius of the earth's orbit.

2.3.2 Unperturbed Orbits

Although we have hardly any interest in planets within the scope of this study, the ideas in this section will be explained with reference to planetary orbits. These ideas will be applied in Sect. 2.3.3 to the theory of the moon with only little changes. They were historically more readily applied in the planetary setting, and they belong to the generally desirable background knowledge about celestial phenomena.[13]

In first approximation, the motion of planets around the sun (and of the moon around the earth) can be treated as a two-body problem, disregarding all other masses. Numerous textbooks exist that teach how to formulate the two-body problem as a set of differential equations, and how these can be solved.[14] The solution is generally formulated in the form of the three Keplerian laws: that the planet revolves in an elliptic orbit, with the sun at one of the foci; that for each planet, the sun–planet radius describes equal areas in equal time intervals; and that for all planets, the ratio of the squares of their periods of revolution to the cubes of the semi-major axes of their orbits is the same.

Thus, let S in Fig. 2.4 be the sun and X a planet. According to Kepler's first law, the planet's orbit is an ellipse $AXX'PA$. Suppose that the planet moves from X to X' in a time interval τ, and that it takes a time T to complete its orbit and return to X. The second law dictates that the area of sector SXX' is to the area of the ellipse as τ is to T. According to the third law, $T^2 : (PC)^3$ is a fixed ratio for every planet. More accurately, this ratio depends on the masses of the sun and planet, but nearly all the mass of the solar system is concentrated in the sun; hence the ratio is nearly the same for all planets (but not for planetary satellites such as the moon), and Kepler's third law is true only approximately.

Nowadays, these three rules are considered the greatest contributions of Kepler to astronomy. Newton showed in *Principia Mathematica* that they had a common cause in the law of gravitation. The same Keplerian laws hold for the motion of the moon around the earth, at least in first approximation, but the moon will be further considered in the next section.

In the case of most planets, the approximation by way of an elliptic orbit is so good, and perturbations so small, that astronomers have adopted a comprehensive apparatus to describe the dimensions and orientation of the ellipse with respect to the plane of the ecliptic. The terminology of this apparatus, which we will introduce next, has for a large part been inherited and adapted from the Greek astronomers who lived with other models of planetary motion, where circles formed the basic building blocks instead of ellipses.

The location of one of the foci of the ellipse is the central body: for planets it is the sun. The dimensions of the ellipse (see Fig. 2.4) are characterized by its semi-major axis $PC = AC$ and eccentricity $SC : PC$. To know the orientation of the ellipse in space, one needs to know the plane of the ellipse as well as the orientation of its major axis within this plane.

The plane of the ellipse is specified by two parameters: its angle of inclination with respect to the plane of the ecliptic and the direction of the line of

---

[13] Moreover, Kepler's third law would make no sense if the discussion were restricted to the lunar orbit alone.

[14] For example Moulton (1902), Brouwer and Clemence (1961), and Roy (1978).

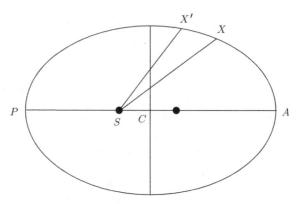

Fig. 2.4 Elliptic orbit

intersection of the plane of the ecliptic and the orbit. This line is specified as the so-called longitude of the *ascending node*, which is the longitude of the point where the planet in its orbit crosses the ecliptic from south to north.

To orientate the ellipse in this plane, it is sufficient to specify the position of one of the extremities of the major axis AP (collectively called the apsides): either the longitude of perihelion P (the position in the orbit closest to the sun) or of aphelion A (where it is furthest away from the sun). In Mayer's time, it was customary to specify the aphelion position for planets and the apogee position for the moon, but the perihelion position for comets, because comets could never be observed near aphelion. The longitude of aphelion is measured from the vernal equinox along the ecliptic to the ascending node and then along the orbit to the aphelion. It is the sum of two consecutive arcs in two planes.

The parameters semi-major axis, eccentricity, inclination, ascending node, and longitude of aphelion are five of the so-called *orbital elements*. The sixth and last orbital element is the position of the planet in its orbit at a specified time.

Keplers laws and the six orbital elements pin down the position of a planet for any past or future moment. The second law of Kepler, also known as the area law, implies that the motion of the planet is not uniform. Instead, it is swiftest at the perihelion and slowest at the aphelion. To arrive at the position of the planet at any specific instance in time, it is natural to relate it to one of the apsides first, and then to use the orbital elements to transform the position in the orbit to a position in ecliptic coordinates.

To ease the computation of the position of the planet in its orbit, one first assumes an imaginary planet that moves with uniform angular velocity and that coincides with the real planet in the aphelion (and, by symmetry, also in the perihelion). The position of this imaginary planet with respect to the aphelion is a linear function of time, and the difference between the imaginary and real planet follows from the area law and the properties of the ellipse.

In astronomical terminology, the imaginary planet represents the *mean motion* and the real planet represents the *true motion*. The angle ASX in Fig. 2.4 is termed the *true anomaly* if X is the real planet or the *mean anomaly* if X is the imaginary planet. The mean anomaly v can be computed as $v = 360° \frac{t}{T}$, where t is the time elapsed since aphelion passage and T is the orbital period of the planet. The difference between the true anomaly θ and the mean anomaly is given by the *equAtion of centre*:

$$\theta - v = -\left(2 - \tfrac{1}{4}e^2\right) e \sin v + \tfrac{5}{4}e^2 \sin 2v - \tfrac{13}{12}e^3 \sin 3v - \tfrac{103}{96}e^4 \sin 4v + \cdots , \quad (2.1)$$

where e is the eccentricity of the orbit. Remember the typographical convention to invoke the astronomical notion of 'equAtion', meaning a periodic correction of mean motion; in addition, note that one equAtion may consist of several terms of a trigonometric series. This particular example is a series approximation suitable for computations; besides, it is the only equAtion of longitude that is involved in unperturbed motion.

2.3.3 The Lunar Orbit

The moon's motion, however, is notably less regular than the motions of the planets. The usual Keplerian laws of elliptical motion fail to account properly for the moon's apparent erratic behaviour, as a diligent naked-eye observer may detect. But Newton's theory of gravitation can equally well account for the perturbations of lunar motion as for the more regular planetary motions. The same force of gravity which keeps the planets in orbit around the sun keeps the moon in orbit around the earth. In the course of its orbit, the distance between the moon and the sun varies, and consequently the gravitational force between sun and moon varies during a month by about 0.5% from its mean value, which is enough to cause considerable trouble.

The moon's motion is not so ill-behaving, though, that the basic model of an elliptical orbit need be abandoned altogether, it only needs some refinement. Basically the same discussion for unperturbed orbits applies to the moon as well, with the understanding that the earth is the central body instead of the sun, therefore the moon's elliptic path has the earth at a focus and the extremities of the apsidal line are termed the apogee and perigee. The concepts associated with an elliptical orbit, such as equAtion of centre, eccentricity, anomaly, etc., can be maintained.

The maximum equAtion of centre of the moon is about $6\frac{1}{3}°$, answering to an eccentricity of 0.055. An ellipse with the same eccentricity and with the diameter of a bicycle tyre exhibits a difference between its major and minor axes that would fit in the tread groove.[15] This illustrates how close the lunar

---

[15] Let a be the semi-major axis, b the semi-minor axis, and $e = 0.055$ the eccentricity. These are related by $a^2 = b^2 + e^2 a^2$; take $2a = 28$ in. (711 mm) and verify that $2b =$

orbit actually comes to a circle. Perturbations in parallax (as a measure of distance) are considerably less observable than perturbations in longitude.

The solar perturbations manifest themselves most visibly as slow changes in the orientation of the idealized elliptical lunar orbit. The rotation of the apsidal line has been mentioned earlier; the apogee completes a revolution in 3,232 days. Additionally, the line of nodes travels backwards around the ecliptic in 6,798 days, or in other words, the orbital plane of the moon rotates once around the axis of the ecliptic in that time. These periods are with respect to the reference point in the ecliptic, viz. the vernal equinox ♈.

The moon itself completes its orbit relative to the equinox in 27.32 days on average, this is the length of the so-called tropical month. With the length of the year about 365.25 days and some basic arithmetic, it follows that the moon returns to the same position with respect to the sun in 29.53 days, this is the so-called synodic month (or lunation) after which the moon repeats its phases. Also, it returns to the same place with respect to its apogee in 27.55 days (this is called the anomalistic month) and to its ascending node in 27.21 days (the draconic month). All these periods are mean values only; deviations occur due to perturbations.

A striking near-commensurability arises between these periods, which is of particular importance to the prediction of eclipses (eclipses occur when the sun and moon are sufficiently close to the nodal line of the lunar orbit). It so happens that 223 synodic months, 239 anomalistic months, and 242 draconic months all are very nearly $6,585\frac{1}{3}$ days, that is 18 years and $10\frac{1}{3}$ or $11\frac{1}{3}$ days, depending on the number of leap days. This period is known as the Chaldaean period, or Saros.[16] After a Saros the geometry of the earth–moon–sun system is very nearly repeated. This implies that if an eclipse occurs at a certain date, then an eclipse will occur again under very similar circumstances after one Saros has elapsed. Moreover, the solar perturbations on the moon's orbit are nearly cancelled out over this period, so that the lunar orbit is almost periodic over one Saros.[17]

The Saros is not only useful for eclipse predictions. For instance, Halley envisaged that an arbitrarily right or wrong set of lunar tables could be used to predict lunar positions with fidelity provided that its error be known for a date exactly one (or perhaps two or more) Saros intervals in the past. For this reason Halley, after he was appointed the Astronomer Royal in 1720 at the age of 65, set out to fulfil an entire 18-year Saros cycle of lunar position observations. However, Halley's idea is only of limited value for longitude finding because the perturbations are not exactly periodic over a Saros.

27.96 in. Alternatively, rework the relation into $2(a-b) = 2e^2a^2/(a+b) \approx e^2a \approx 0.04$ in, or 1 mm.

[16] The name 'Saros' was given to this time-span by Edmond Halley in 1691. Although the Babylonians were familiar with it and used it for eclipse predictions, they did not use that name for it.

[17] Roy (1978, p. 285) and Perozzi et al. (1991).

Reversely, Mayer used the Saros periodicity whenever he met an exceptionally large difference between an observed and a calculated position of the moon, in order to verify whether the observation or the calculation was faulty. Illustrative is also his *Catalogus Eclipsium Lunae*, a systematic comparison of lunar eclipse observations and predictions ordered according to the Saros principle.[18]

Thus far, we have seen that the moon's orbit can be approximated as an ellipse that slowly changes its orientation. The moon's position in such an orbit can be described by a set of *mean positions* and *mean motions*, which together specify the positions of the mean moon, mean apogee, and mean node as linear functions of time, with the adjective *mean* to signal that periodic terms are not taken into account. We have also met with the secular acceleration, which is partly of an extremely long period, partly dissipative, and which can be modelled effectively by adding a quadratic term to the mean longitude. Next, the various eqAtions are taken into account, of which the lunar equAtion of centre is the largest.

We now turn to a brief historic overview of some other equAtions of the motion of the moon. As far as longitude is concerned, all equAtions consist of (sums of) terms of the form $c_k \sin k\alpha$ ($k \in \mathbb{N}$), where α is known as the argument and the c_k as coefficients.

Hipparchus (ca. 150 BC), building on the work of those before him, had developed a rather accurate model for solar and lunar motion. In his model, sun and moon went about the earth at constant speed in eccentric circular orbits, i.e., in circles whose centres did not coincide with the centre of the earth. The model incorporated the inclination of the lunar orbit to the ecliptic, the mean motions of the lunar apsidal and nodal lines, and his greatest discovery, the precession of the equinoxes. His model could predict lunar and solar eclipses reasonably well, but his observations showed discrepancies for lunar positions away from the phases of the New and Full Moon.

Ptolemy, ca. AD 150, constructed an elaborate theory[19] to account for the most prominent discrepancies of Hipparchus' theory. Ptolemy's model is successful in predicting lunar longitudes, but has a very unsatisfactory consequence of varying the earth–moon distance and hence also the apparent lunar diameter by a factor of nearly two. This was recognized and corrected independently by Ibn al-Shāṭir and Copernicus. Centuries later, the equAtion that Ptolemy had addressed in order to correct the position in the quadrants (i.e., near the first and last quarters) was named *evection* by Bulliau. It is an equAtion which Jeremiah Horrocks in 1638 successfully modelled as a variable eccentricity of an elliptical lunar orbit. His invention will be the subject of

---

18 The *Catalogus* was appended to Mayer (1754).

19 'Lunar theory', or theory in general, has long been the name for what we would prefer to call a (kinematic) 'model' nowadays. In the words of Francis Baily, in the seventeenth century it meant 'rules or formulae for constructing diagrams and tables that would represent the celestial motions and observations with accuracy' (Baily 1835, p. 690), quoted in Newton (1975, p. 3).

Sect. 6.3. Evection, being the second largest of the lunar equAtions, may amount to $1\frac{1}{3}°$. In the Mayerian notation that we adhere to, its argument is denoted as $2\omega - p$.

Further irregularities were discovered by Tycho Brahe shortly before 1600. The third equAtion in size and chronology is the *variation*, which attains a maximum of some $\frac{2}{3}° = 40'$. Its argument is 2ω, which is twice the angular separation of the sun and moon, hence it vanishes at the syzygies and quadratures.[20] One of the greatest successes that Isaac Newton achieved in *Principia* was his complete accounting for the variation on the basis of gravitation. Variation is principally caused by the veering of the solar direction of pull while the moon completes its orbit around the earth; in particular, it is independent of the lunar eccentricity.

Tycho and Johannes Kepler independently discovered the *annual equAtion*, which Newton later understood as being caused by the variable distance of the sun from the earth–moon system; it is brought about by the eccentricity of the earth's orbit. It has the solar anomaly ς as argument and reaches a maximum of slightly more than $11'$. Tycho also discovered the two prime equAtions of the lunar node and inclination. Kepler's was the first lunar theory to incorporate the idea of an elliptic orbit; his theory was, however, not very successful.

All these equAtions had been discovered observationally. Newton was the first to deduce new equAtions from theory, truly an outstanding feat in itself. With the refinement of mathematical methods since the eighteenth century, more and more new equAtions were derived, and the practice of baptizing equAtions was soon dispensed with.

One equAtion that *did* receive a name is the *parallactic equAtion* caused by the variable moon–sun distance over the course of a month. This parallactic equAtion depends on argument ω, making for a convenient combination with the above-mentioned variation when tabulated. By its nature, this equAtion provides a link between the earth–moon distance and the moon–sun distance, and hence between the lunar and solar parallaxes. The lunar parallax is quite large, and by the middle of the eighteenth century it was feasible to measure it almost directly: the Abbé Nicholas Louis de Lacaille was then temporarily observing at the Cape of Good Hope well to the south of Europe, while Lalande was sent to Berlin, nearly on the same meridian, and between them they had a sufficiently long baseline.[21]

The solar parallax, on the contrary, happened to be one of the most sought after, yet elusive, constants in astronomy. Its value was crucial not so much because our knowledge of the size of the solar system depended on it, but because its erroneous value propagated into the orbital elements of all the observed solar system bodies. Two transits of Venus before the sun, first in 1761 and again in 1769, were carefully observed from various places on

---

[20] Syzygy: the collective appellation of Full Moon and New Moon; quadrature: idem for First and Last Quarter.

[21] See Chapin (1978, p. 167) for further details.

earth in order to deduce a value for the solar parallax. Unfortunately this subject, of the highest interest in the history of science, is beyond the scope of the current work. Let it suffice to remark that the link between the two parallaxes provided by the parallactic equation was signalled by Leonhard Euler, and exploited by Tobias Mayer, who obtained, in 1755, a value for the solar parallax of 7.8″. His result is comparable with the yield of 8.5″ to 10.5″ produced after the Venus transit expeditions. The modern value of solar parallax is approximately 8.8″.

To end this crash course in lunar theory, it may be well to point out that the difficulty of lunar theory lay not only in the complexity of the mathematics of the three-body problem, but also in the application of observed positions. It had to be taken into account that observations contain measurement errors, some of which could be dealt with approximately (such as atmospheric refraction), while others were inherently unknown (such as random observational errors). Astronomers were among the first to realize this in a general sense and to develop methods to deal with observational errors. Moreover, it is no easy matter to determine for instance the position of the lunar apogee from observations: not only because the radius vector changes its length slowly within narrow limits, but also because all the other small inequalities are superimposed on the elliptic motion. 'A major obstacle... lay in the multitude of small inequalities, which they had no way of beginning to discern,' wrote Curtis Wilson.[22]

We can see mathematics at work in several ways. First of all, there is the mathematics of circles and ellipses which are used to construct kinematic models of motion. Then there is also the mathematics of Newton's *Principia*, later also the mathematics of differential equations and trigonometric series, not to mention Hamiltonian dynamics, forming the tools of *physical astronomy*, or *celestial mechanics* as it came to be known since Laplace. Finally, there is the mathematics of the 'combination of observations', of inferring appropriate values for model parameters from data. The affiliation between astronomy and mathematics is indeed long and intense.

---

[22] Wilson (1989a, p. 196).

Chapter 3
The Pioneer's Work

Although this is not the place or time for a comprehensive biography of Tobias Mayer, a short account of his – likewise short – life is desired here.[1] My narrative is biased towards the facets of Mayer's life (1723–1762) that are relevant to the current investigation of the development of his lunar tables. Therefore, almost everything related to the practical pursuit of astronomy is left out, including what pertains to instruments, observing, or the reduction of observations. On the other hand, this chapter contains several new results concerning the development of the lunar tables. These results are worked out in more detail in the ensuing chapters, to which I will refer where appropriate. The current chapter, devoid of most of the technical detail, may also serve as a backdrop against which the more technical topics in this book will fall into perspective.

3.1 Youth and Early Career

Tobias was born in Marbach am Neckar in 1723, and raised in Esslingen, both towns close to Stuttgart in the German state of Württemberg. His parents died while he was still a boy, but fortunately for him his talents were recognized, and he was enabled to attend Latin school, meanwhile developing his early interests in drawing, fortification, and cartography. Through these interests, he came into contact with geometry. Mathematics was not a subject taught in his school, but Mayer found company with a local shoemaker with a similar interest. The shoemaker had money to buy mathematical texts but no time to read them, therefore they agreed that Mayer read the books and taught their contents to the shoemaker mending shoes. After mastering Christian von Wolff's *Anfangs-Gründe aller mathematischen Wissenschaften*

---

[1] See Forbes (1980) for a more balanced biography, from which most biographical information in this chapter is taken.

S.A. Wepster, *Between Theory and Observations*, Sources and Studies in the History of Mathematics and Physical Sciences, DOI 10.1007/978-1-4419-1314-2_3,

and similar works, Mayer wrote his first treatise, published in 1741, on a new and general method for the inscription of polygons in circles, a subject clearly displaying his interest in the theory of fortifications.

Mayer's attempted career with the military did not materialize. Instead, he found employment in Augsburg with the publisher Johann Andreas Pfeffel in 1744. The next year saw the publication of Mayer's impressive *Mathematischer Atlas*, in which the then 22-year old author provided an overview of the mathematical sciences in 60 plates, divided over the subjects of arithmetic, geometry, trigonometry, geography, chronology, gnomonics, fortification, artillery, architecture, optics, and mechanics.[2] Not long after, Mayer moved to Nuremberg, a town that could look back on a long astronomical heritage since the days of Regiomontanus and Bernhard Walther in the late fifteenth century. Here, with his *Mathematischer Atlas* providing credentials, Mayer entered the prestigious geographic firm of Johann Baptist Homann's heirs, which held a leading position in German cartography as well as high ambitions for the future.

A very fruitful period in his life began with Mayer joining in to draw several maps for the *Gesellschafts Atlas*, which the Homann house published in 1747. Soon Johann Michael Franz (one of the firm's directors), Tobias Mayer, and his colleague Georg Moritz Lowitz, were pursuing the plans of their own Cosmographical Society, aimed at the (re-)structuring of German cartography, no doubt also with an eye on the increasingly difficult cash position of the main firm.

By way of illustrative depiction of the accuracy – or rather the lack thereof – with which the latitudes and longitudes of German places were known, Mayer composed *Mappa Critica*,[3] a map showing three Germanies dislocated according to three geographers: de l'Isle (probably Guillaume, 1665–1726), Homann, and Mayer himself. The latitudes of the charted cities were mostly consistent, whereas the longitudes of some places (such as Dresden or Prague) differed by as much as a degree, thus showing to what extent longitude was problematic even on land. Mayer drew on a wide variety of sources and was thus confronted with the conflicting data provided by the various reports that he employed. The handling of such data sets will be seen to be a recurring theme in his work.

Other projects for the Cosmographical Society included the production of lunar globes and the dissemination of research results of its members in *Kosmographische Nachrichten und Sammlungen*, which was intended as a series with volumes to appear annually. Mayer contributed to these goals with a comprehensive programme of mapping the visible lunar surface, for which a careful quantification of lunar libration was necessary: here is another example of handling large amounts of data, which we will further explore in

---

[2] See Mayer (1745).

[3] Mayer (1750c).

Chap. 9. For diverse reasons, the Cosmographical Society never produced a single lunar globe, but recently the project has attained a new life.[4]

The intended series of the *Nachrichten* was rather short-lived, because only one volume was produced.[5] Characteristically, in his numerous contributions to the volume, Mayer was keen to show his careful handling of astronomical observations and his ability to procure everything possible from them. He was now making astronomical observations both from the premises of the Homann firm and from the observatory of the former astronomer Eimmart – although the latter place was in such a state of neglect that he had to bring a hammer to operate the sector. In Mayer's eyes, astronomy was at the service of geography:

> It is only too well known in geography that many, indeed most, of the numerous known and noteworthy places need to have their latitude and longitude more accurately determined. To achieve this is one of the most important objects of astronomy [...].[6]

As Forbes showed, almost all of Mayer's further professional endeavours were guided by this assessment.

3.2 The Geographer's Lunar Tables

Finding the longitude of places was greatly assisted by proper knowledge of the moon's motion, and therefore the moon became an object of interest to Mayer. It is evident from various sources that he used Leonhard Euler's lunar tables of 1746, and the almanacs in the *Berliner Kalender*, which were based on those tables, in the period when he was working at the Homann cartographic office in Nuremberg.[7] Mayer had his own manuscript copy of Euler's lunar tables, to which we will refer with the alias duin from now on.[8] In Sect. 9.5, I discuss an attempt of Mayer's to adjust the coefficients in the duin tables to observations. Apparently, he was not satisfied with Euler's tables, and he was eager to construct tables of his own.

---

[4] A number of sectors for the globe were engraved in copper during Mayer's lifetime. These sectors have been preserved and the missing sectors have recently been engraved, based on Mayer's drawings of a lunar map. A limited number of globes are now being produced, working from authentic prescriptions. This project is initiated by the *Tobias Mayer Museum Verein* in Marbach am Neckar.

[5] Mayer et al. (1750).

[6] Quoted from (Forbes 1980, p. 42). The original is in *Untersuchungen über die geographische Länge und Breite der Stadt Nürnberg* (researches into the geographical longitude and latitude of the town of Nuremberg), Cod. $\mu_{11}^{\sharp}$, reprinted in Forbes (1972, I, pp. 33 ff.).

[7] For example, Mayer et al. (1750), correspondence of Franz to Euler, (Juškevič 1975, pp. 119, 120). See Forbes (1980, pp. 55, 56). The tables are Euler (1746).

[8] See the aliases list in Appendix A. Mayer's copy of Euler's tables is in manuscript Cod. $\mu_{14}^{\sharp}$.

In the beginning of 1751, when he was still in Nuremberg, Mayer proudly announced to the astronomer Joseph Nicholas De l'Isle (1688–1768), with whom he was in regular correspondence, that he had produced new lunar tables of his own. He averred that these (alias kreek) were accurate to about 2′. He added that these tables were easy to use.[9] I have been unable to locate the kreek tables among Mayer's manuscripts, nor even an indication of their provenance. However, I found several calculations of the moon's position in Mayer's manuscripts[10] that match his description of the tables. It is almost certain that Mayer employed the kreek tables for these calculations.

3.3 New Opportunities

The ambitious Cosmographical Society of Nuremberg attracted the attention of the young and expanding Georg-August Academy in Göttingen. The academy hoped to be able to attract the Society in order to further its own goals. Therefore, Mayer was offered a professorship in practical mathematics together with the responsibility (to be shared with Johann Andreas Segner) for a new astronomical observatory. Plans for the observatory had developed after George II, King of Great Britain and Elector of Hanover, had visited the academy in 1748.

---

[9] 'I have just finished a very long and painstaking algebraical calculation that I have undertaken on the Lunar Theory in the Newtonian system. In it, I have deduced Lunar Tables that are much more exact than any others, because they give the true place of the Moon to about 2 minutes precision, as I have recognized after having compared them with more than 20 observations made in different aspects. They are not less convenient for the calculation, because one does not need to fetch first of all the true place of the sun, nor the true or eccentric anomalies, as in those of other Astronomers [such as Euler or Lemonnier]. It suffices to take the mean motions in order to find in the Tables the equations, which are at the same time all additive, so that the reduction of the mean place of the Moon to the true place happens at the same time & through a single addition'. («*Je viens de finir un calcul algebrique fort long et penible, lequel j'avois entrepris sur la Theorie de la Lune dans le systeme Newtonien. J'en ai deduit des Tables Lunaires beaucoup plus exactes qu'aucunes autres puisque elles donnent le vrai lieu de la Lune à 2 minutes pres, ce que j'ai reconnu en les ayant comparées avec plus de 20 observations faites dans differents aspects. Elles ne sont pas moins commodes pour le calcul. Car on n'y a besoin de chercher d'abord ni le vrai lieu de soleil ni les anomalies vraies ou excentriques, commes dans celles des autres Astronomes. Il suffit de prendre les moyens mouvemens pour trouver dans les Tables les équations, qui en mêmes tems sont toutes additives, de sorte que la reduction du lieu moyen de la Lune au lieu vrai se fait à la fois & par une seule addition.*») So far, he has made mention of Newtonian theory only, not of fitting his tables to observations, but after having promised to send the tables he asked if De l'Isle would be willing to communicate some of his best lunar observations 'in order to be able to further improve my Tables before publishing them' («*pour être en état de pouvoir perfectionner encore plus mes Tables avant de les publier*»). Mayer to De l'Isle, Nuremberg 14 Jan. 1751, published in Forbes (1983). See also (Forbes 1972, I, p. 15); however, most of Forbes' assertions in his introductory chapter must be rejected.

[10] Cod. $\mu_6^\sharp$ and the first folios of Cod. $\mu_{41}^\sharp$.

Mayer accepted the offer and arrived in Göttingen around March 1751, four years later followed by his colleagues of the Cosmographical Society, Lowitz and Franz, who had also been persuaded to take up positions at the academy. Although financial problems shattered the Cosmographical Society after its relocation into the Hanoverian state, Mayer's research took off on a new footing. The emphasis in his work had already been drifting away from cartography towards astronomy, and in his new academical surroundings an increasing demand was placed upon his research and teaching abilities.

In July 1751, soon after taking up his professorship in Göttingen, Mayer started a correspondence with Euler who was then in Berlin. Euler had written several memoirs concerning orbital motion, including one on the great inequality of Jupiter and Saturn; he had also prepared the duin lunar tables already mentioned, from which ephemerides for the *Berliner Kalender* were computed. Thirty-one of the letters they exchanged have been preserved.[11] They provide a most valuable insight in many aspects of Mayer's scientific work and in the relation between Euler and Mayer. The topics that were discussed included refraction of light in the atmosphere, parallax of the moon, theories of gravitation and ether, lunar theory, and many more.

In his first letter, Mayer expressed his admiration for Euler's diverse treatises on celestial mechanics, especially the treatise on the great inequality of Jupiter and Saturn,[12] from which Mayer had learnt some essential mathematical techniques. Euler's treatise had won a contest of the Paris Academy, which was held annually in consequence of the will of Rouillé de Meslay, as mentioned above on p. 13. Considering the influence of Euler's treatise on Mayer, we see that the will of Rouillé de Meslay indeed had a significant impact on the determination of longitude at sea. Mayer applied the techniques that he had learnt through Euler's publications in the derivation of his own lunar theory from the law of gravitation, to be discussed in Chap. 5. However, he admitted that he was not completely successful, and I will show (Chap. 6) that his theory remained incomplete and in a problematic state until 1755.

In that same first letter to Euler, Mayer referred to a method of predicting lunar positions that had been put forward by Edmond Halley. This method exploited the fact that the mutual positions of sun, earth, and moon in their orbits are nearly repeated after a Saros – and hence also the error pattern of an arbitrary set of lunar tables. Mayer, in his letter to Euler, quoted his own formula with which he could predict a future lunar position, given a position one or more Saros periods earlier:[13] in other words, his formula expressed the deviation of the moon's position from full periodicity. This seems to be the only reference of Mayer to the predictive use of the Saros periodicity. I have not been able to find computations among Mayer's manuscripts that are connected to either the derivation or the use of the above-mentioned

---

[11] The letters were published in Forbes (1971a); see fn. 4 in Chap. 1.

[12] See p. 10.

[13] Forbes (1971a, pp. 34, 35).

formula. In later years, Mayer employed the Saros periodicity (but not the formula) to select pairs of observations of near-identical lunar positions, with which he could fit the mean motion coefficients in his lunar theories. In other cases, such a pair could indicate whether his tables or a faulty observation were to blame for an unusually large discrepancy.

Almost a year later, in January 1752,[14] Mayer reported to Euler a rather different set of lunar coefficients, to which I assigned the name zand. He made no mention any more of the former arrangements of the kreek tables, nor of Halley's Saros method; yet all the arguments of the set were still combinations of the mean motions. Mayer said that some of the coefficients had been adjusted to observations, but it is unknown to us how he accomplished that. I found traces of this stage of his work on lunar theory in several of his manuscripts, some of which I will present in the following chapters.

By that time Mayer had also finished an investigation of lunar parallax. In a memoir on that subject, he warned that parallax could not be accurately determined without paying attention to the oblate shape of the earth, and he explained two independent ways to arrive at lunar parallax: one way exploited a relation between the orbital period of the moon and the (lengthened) period of a pendulum at the equator (the distance to the centre of the earth governs both), the other depended on the rapidly changing effect of parallax on the observed position of the moon when the moon approaches the meridian.[15]

Forbes asserted that: 'Mayer's recognition of the need to take account of the Earth's spheroidal shape now involved him in the calculation of a large number of new inequalities, of which he retained sixteen...'.[16] Indeed Mayer wrote to Euler that he had considered the influence of the earth's oblateness; I even located Mayer's computations to that effect in Cod. $\mu_{30}^{\sharp}$, together with some of Mayer's work on atmospheric refraction. But the sixteen terms that Forbes alluded to are simply too sizable to be the result of the earth's

---

[14] Forbes (1971a, p. 48).

[15] The essence of the second procedure is this: let the moon be observed at two known times, say an hour apart. Let $\alpha_1 - \alpha_2 = \Delta\alpha$ be the observed difference of the moon's right ascensions at those times, and let $\alpha_1' - \alpha_2' = \Delta\alpha'$ be the difference of right ascensions at these times according to lunar positions computed from tables. The parallax in right ascension is $P \cos\varphi \sin\beta/\cos\delta$, where P is the unknown equatorial lunar parallax, φ is the latitude of the observer (corrected for oblateness), β is the apparent azimuth of the moon, and δ its declination. Ideally, the parallax at any of the two times would equal $P = \pm(\alpha_i - \alpha_i')$, but the tabulations for α_i' were rather inaccurate. However, the *velocity* of the moon was much better represented in the tables, at least for relatively short time-spans, and so Mayer could circumvent the inaccuracy of the tables: the difference of true and observed displacements equals the difference of parallax in right ascension, or

$$\Delta\alpha' - \Delta\alpha = P \cos\varphi \left(\frac{\sin\beta_1}{\cos\delta_1} - \frac{\sin\beta_2}{\cos\delta_2} \right),$$

in which P is the only unknown. In this memoir, Mayer (1753a), Mayer still used Euler's lunar tables from the *Berliner Kalender*.

[16] Forbes (1980, p. 136); see Forbes (1971a, pp. 10, 47–49) for the letter to Euler referred to in the text.

oblateness. In fact, they are just terms of the zand version of lunar tables mentioned above. Mayer wrote their coefficients in a letter to Euler in a passage following his statements on the determination of the lunar parallax; hence, I presume, Forbes' misinterpretation. Mayer's later report to Euler that he had 'completely and utterly ignored many small equations, which are similar to those which depend on the shape of the Earth, since they scarcely amount to $12''$ or $15'''$,[17] suggests only that he was aware of the existence of small oblateness inequAlities in lunar longitude.[18]

3.4 A Shift in Strategy: Multiple Steps

In January 1753, Mayer reported to Euler again a completely new version of tables, derived, as he says, from theory and adjusted to observations.[19] Four successive steps are needed to compute the position of the moon with these tables. In each step, some arguments with which the tables are entered are modified in a specific way: the arguments are no longer equal to the mean motion arguments, as they were before.

This 'shift in strategy',[20] which seems to fall out of the blue, leads us now to distinguish the *multistep* form of these tables, formulae, and computations where arguments are modified in several steps, from the former *single-step* ones that use only mean motion arguments throughout. A more thorough explanation of these two forms is contained in Sect. 4.1 and, more specifically, Sect. 4.2.

Mayer had checked his new tables with more than 200 observations, and found only ten or twelve that deviated about $1\frac{1}{2}'$. Mayer mildly adjusted these tables, and then published them (henceforth designated as kil) a few months later in the proceedings of the Göttingen Scientific Society.[21] Among Mayer's manuscripts, I have found new information showing that Newton's lunar theory of 1702,[22] as represented by Lemonnier's lunar tables,[23] had a profound impact on kil, in particular on Mayer's shift of strategy. This impact is the subject of Chap. 6.

---

[17] Forbes (1971a, p. 63).

[18] Mayer's tables after 1754 included an *empirically* detected equAtion, with argument the angular distance of the moon from the ascending node of its orbit, of only $4''$. Laplace later showed that this equAtion is caused by the oblateness of the earth (Laplace 1802, pp. 658–659).

[19] Forbes (1971a, pp. 61–63).

[20] Quoted from Forbes and Wilson (1995, p. 64).

[21] Mayer (1753b).

[22] See Newton (1975). Newton's 'lunar theory' of 1702 held a prescription, or algorithm, for the computation of lunar positions; it may be a surprise to the modern reader that there was no explanation by way of a theory of gravity in it.

[23] Lemonnier and Keill (1746).

In this period Mayer also created a new tool to help improve the accuracy of the tables. This tool has not been recognized and described in earlier research. It is of high importance for our understanding of Mayer's work on lunar tables and of the early development of statistical ideas. We will study it fully below in Sect. 8.5. The tool embodied a means to keep track of the effectiveness of various amendments that Mayer tried to the coefficient values. The sheets of paper on which he systematically recorded these effects show a remarkable similarity, both in form and function, with the modern objects that we commonly refer to as spreadsheets. In absence of any indication of what Mayer called them, I will not hesitate to apply the anachronistic but apt term *spreadsheets* to these papers, or *spreadsheet tool* when the emphasis is on their functionality.

3.5 Contending for the Longitude Prize

Because of the high accuracy of the published kil tables, the determination of longitude at sea by lunar distances came within reach. Prior to this development, references to longitude at sea are strikingly absent from Mayer's correspondence and notes, yet without the slightest doubt he must have been fully aware of the problem and of the rewards offered for a solution. Being rather modest about his own accomplishments, Mayer was encouraged by Euler, and urged by Michaelis, to submit the kil tables to the Board of Longitude in Great Britain, to compete for the Longitude Prize stipulated by the Act 12 Queen Anne. The philologist Johann David Michaelis (1717–1791) was a key figure in the events to be described below. He was secretary of the Göttingen Scientific Society and editor of its proceedings in which the kil tables were published. Besides, he was secretary for Hanoverian affairs in Göttingen. Because his cousin William Philip Best was one of the private secretaries of King George II, a diplomatic channel to London existed through Michaelis. Through this channel, communications on Mayer as a candidate for the Longitude Prize were started in September 1754.

A few months earlier, the Göttingen *Commentarii* containing Mayer's kil tables had already reached the other side of the North Sea, where they were received with guarded enthusiasm. A preliminary check of the tables against observations showed one or two exceptionally large differences, which were soon traced to errors in the reduction of the observations. An extract of Mayer's own preface to the tables was published in the *Gentleman's Magazine*.[24] As Forbes pointed out, Mayer himself was somewhat reluctant to approach the Board of Longitude, because he realized that the Board would

---

[24] The *Gentleman's Magazine* for August 1754, contains a translation into English of a substantial part of Mayer's Latin preface, signed 'B.J.' Forbes included most of the English article in his biography of Mayer and identified the author as John Bevis (Forbes 1980, pp. 143–146). Section 6.8 below discusses Mayer's preface.

not consider whether his sole *tables* were within the accuracy requirement, but that the Board rather required a complete *method* that could stand up to the accuracy requirements in an actual trial at sea.[25] So Mayer needed first to embed his tables in such a method, and second, he had to improve their accuracy, because when in use at sea, both the astronomical observations and the ensuing calculations would further dilute the overall accuracy of the method.

Thus, Mayer continued to refine his tables by observations and the spreadsheet tool. Previously, he had employed timed lunar meridian passages and lunar eclipses. But since his published kil tables were already of about the same accuracy as those observations, improvement was no straightforward task. To proceed, Mayer selected observations of higher quality, namely those of occultations. At an occultation, the moon visually intercepts the light from a star or planet. The moon covers and uncovers the other body almost instantaneously; therefore, occultations can provide accurately timed positions of the moon relative to the occulted body.[26]

Mayer was well aware of the inaccuracy of stellar positions in the star catalogues then in use, and therefore he started out with occultations of only one star, Aldebaran. Any catalogue error in Aldebaran's position would then only be reflected in the lunar epoch position. To fix the epoch position, and consequently to correct the coordinates of Aldebaran, he used solar eclipse observations. Then, gradually, he admitted occultations of other stars into his data set.[27]

In the same period, Mayer was negotiating his resignation from Göttingen to accept an offer of a position in Berlin, where he would be working together with Euler, who had been most instrumental in making a position available for his esteemed correspondent. As it happened, though, Mayer's resignation was refused by nobody less than King George II, who was fully aware of the chances Mayer stood in regard of the Longitude Prize. Mayer's leave to Berlin would not only break the attempts to concentrate the Cosmographical Society in Göttingen – it could also mean a serious loss of prestige. In return, Mayer was in a position to request a salary raise as well as the resignation of Segner, his superior with whom he had to share the observatory, and with whom relations were in a particularly bad shape. The observatory, meanwhile, was in its last stages of completion and had still to be equipped with suitable instruments. Its most important instrument was to be a 6 foot mural quadrant made by the London instrument maker John Bird, who had recently constructed a similar quadrant of 8 feet for James Bradley at the Royal Greenwich Observatory. Bird's quadrant was installed in Göttingen in the winter of 1755–1756.

---

[25] Forbes (1980, p. 153).

[26] Mayer interpreted the suddenness of the occultations as evidence against a lunar atmosphere, and published his then still much disputed opinion in Mayer (1750d).

[27] Mayer reported this procedure to Euler; see Forbes (1971a, pp. 80–84), and Forbes' comment on p. 15, *ibid.* Concurrently, he had derived the moon's latitude to a higher accuracy than anyone before.

Meanwhile, Mayer had prepared a new instrument (now known as repeating circle) of his own invention, and a memoir on how to find longitude at sea with it and his lunar tables. The memoir, together with a description of the instrument, was transferred to William Philip Best in November of 1754.[28] Best sought the advice of James Bradley (the Astronomer Royal, successor of Halley), Lord Anson (First Lord of the Admiralty), and George Parker, 2nd Earl of Macclesfield (president of the Royal Society), all of them ex officio members of the Board of Longitude. Lord Anson must have held the most emphatic ties to the longitude problem of this threesome, since he had lost more than seventy of his crew searching for potable water and the island of Juan Fernandes in the wrong longitude. Mayer received a reaction of the Board members (principally Bradley) via Best and Michaelis:

> But just as Prof. Mayer adduced that he has further improved the lunar tables in the second volume of the *Commentarii* and figured them out nearer; so they regard it necessary that such calculations, and at the same time the principles upon which they took place, are also shown, ere and before the request will be applied to the Admiralty.[29]

The improved tables that Bradley had desired reached Best before Christmas.[30] These tables, rak for short, have never been published, although apparently plans in that direction existed.[31] Nevil Maskelyne retrieved the underlying coefficients of the tables and published them later in the Nautical Almanac for 1774; the manuscript tables themselves are apparently not anymore in the Royal Greenwich Observatory archives now kept in Cambridge and must be considered as lost.

Forbes has made mention of the improvements, but he was unaware of Mayer's techniques. Bradley extensively compared Mayer's tables to almost 1,200 lunar observations until 1760, finding no differences in excess of $1\frac{1}{4}'$.

---

[28] These two were prepended to the edition of the last manuscript tables in Mayer (1770), as *Methodus longitudinum promota* with several additions. The memoir Mayer (1754) is a much more general exposition of the usefulness of his tables for the longitude problem, intended for a broader audience; in it, Mayer announced a second part in which he intended to explain their application to that end. No doubt the *Methodus* is that part. Both these treatises had been intended for publication in the annual Göttingen *Commentarii*, and the withdrawal of the second part in view of the quest for the Longitude Prize led to a quarrel with the printer and disruption of the series.

[29] *"Gleichwie aber der Hr. Prof. Mayer in seinem Manuscripte angeführt, daß er die in dem zweitenn Bande der in der Commentarien befindliche Mondstabellen annoch verbessert und näher ausgerechnet habe; also halten sie nöthig, daß auch sothane Ausrechnungen, und zugleich die Principien, wornach selbige geschehen, angezeigt werden, ehe und bevor das Gesuch bey der Admiralität angebracht wird"* Michaelis (1794–1796) Best to Michaelis, 19 November 1754.

[30] Michaelis (1794–1796) Best to Michaelis, 24 December 1754.

[31] See Forbes (1971a, pp. 96, 114); the London printer Nourse had asked permission to print Mayer's new tables and apparently Mayer agreed.

The paucity of Bradley's observations of the spectacular return of Halley's comet in 1758–1759 is probably due to his elaborate checking of Mayer's tables.[32]

Contrary to Mayer's prompt response to Bradley's request for the improved tables, and repeated requests notwithstanding, it took Mayer almost a year to prepare an account of the '*Principien*' on which his tables were founded. Bradley's request embarrassed Mayer, because his tables were founded on a mix of theory and observations; his use of observations to adjust the coefficients lacked a theoretical foundation. To make matters worse, his tables embodied a multistep procedure that he had adapted for practical reasons from a kinematic contraption in Newton's 1702 lunar theory, for which Mayer had no theoretical justification (this is the subject of Chap. 6). The difficulty to reconcile the theory with the multistepped computational structure of the tables may well have been the cause of his delay. As I will show in Chap. 7, the multistepped structure in itself adds little or nothing to accuracy, but Mayer was probably not aware of that. In consideration of the lack of justification of the multistepped procedure, it may well be doubted whether Mayer had more than an incomplete sketch of a theory when Bradley's request reached him. Mayer's account of the theory and the circumstances of its composition will be further discussed in Chap. 5.

3.6 His Later Researches

After finally completing the theoretical tract, and sending it off to London in the fall of 1755, Mayer continued his efforts to improve the tables using his spreadsheet tool, although not as many table versions were produced as before. He also applied the same method set forth in his lunar theory to the computation of the orbits of Mars and Jupiter, and asserted that his theory of Jupiter was accurate to about 1′.[33]

The quadrant that Bird had fabricated was installed at the new observatory in the winter of 1756. As soon as that was done, Mayer set out to examine the instrument and to register any remaining errors in it and its mounting. His next project was to use the quadrant to measure the positions

---

[32] Arthur Alexander suggested this in his introduction (p. ix) to the 1972 reprint of Rigaud (1832). Bradley's comparisons were archived as RGO 3/33, and they inspired Bradley to improve Mayer's rak tables.

[33] Curiously, Mayer wrote about his Jupiter success to Euler in February 1755, before the delay of his lunar theory was evident and long before its completion. It is not to be expected, however, that his Jupiter theory accounted for the great inequality of Jupiter and Saturn other than through an empirical equation (perhaps a secular one). The research into the Mars orbit may be dated a little later, for he read a memoir on that subject to the Academy in April 1756. Wilson (1985, p. 66) mentions that the *Connoissance des Temps* for 1763 contained Jupiter tables that were based on Mayer's formulae, but I have not verified his claim.

Fig. 3.1 Tobias Mayer

of almost 1,000 stars in the zodiac, and to prepare a catalogue of them. The necessity of such an undertaking must have appeared to him when he was using occultations of stars by the moon to improve his lunar tables, because then he had found significant inconsistencies in the tabulated positions of the stars according to Flamsteed, Rømer, and others. He completed the project, but the catalogue was only published posthumously.[34]

It was by that time well enough known that the stars are not as fixed with regard to their mutual positions as had before been supposed; Halley had announced their proper motion in 1717. By a comparison of his own star positions with the well-documented data of Ole Rømer of half a century earlier, Mayer hoped to be able to detect a pattern in the proper motions from which to conclude the direction of motion of the entire solar system, just like a man walking through a forest perceives the trees as moving, in his own metaphor. He could not decisively discern such a pattern, but the idea was later picked up and developed further by William Herschel and others.

Of Mayer's further research, his investigation of the variation of the thermometer must be mentioned, which we will return to in Sect. 9.2. Other investigations included a theory of colour mixing, of earthquakes (of interest after a quake shook Germany on 18 February 1756, potentially upsetting the

---

[34] The catalogue was included in Mayer (1775), as was a treatise on the investigation of the quadrant. The catalogue attracted attention until at least the end of the nineteenth century, witness Auwers (1894).

mounting of his new mural quadrant), and of magnetism.[35] Mayer regularly read reports of his research to the Academy; abstracts were published in the *Göttingische Anzeigen von gelehrten Sachen.*

Meanwhile, Europe was engaged in one of its bitterest fights, the Seven Years' War (1756–1763); since the summer of 1757, Göttingen was on and off occupied by French troops. Conditions of living and working in the city deteriorated. The astronomical observatory was built on top of a tower which happened to be now also used as a powder-magazine. A similar tower blew up killing seventy people, whereafter Mayer reduced his habitual nightly visits – with a lantern in his hand – to his working place, in all likelihood convinced that it was his task to *observe* the stars, and not to join them. But performing his other duties was not easy either, with officers lodging in his house and lecture halls full of army supplies.

His death came in February 1762, just after his 39th birthday, not by sudden explosion but after a prolonged illness. He left a wife and four children, who lived under a high mortgage in a half-demolished house in a nearly bankrupt city. Unfortunately for them, the Board of Longitude had still not reached a final conclusion regarding either Mayer's claim for the Longitude, or John Harrison's almost simultaneous claim based on his timekeeper.

3.7 The Board's Decision and Beyond

One of Mayer's last wishes was that the latest version of his lunar tables (alias rede) be sent to the Board of Longitude, together with the results of sea trials of his method by his pupil Carsten Niebuhr that had reached him just before his death.[36] These papers were indeed forwarded more than a year later.

Meanwhile, the Admiralty had ordered Captain Campbell to try Mayer's method of longitude determination at sea. Campbell disapproved of Mayer's repeating circle as being too cumbersome to operate, consequently the marine sextant was developed as a Hadley octant with an extended arc and engineered to a higher degree of precision. Although Campbell left the necessary lengthy calculations for finding the longitude to Bradley on shore, the tests showed that longitude could in principle be found.

---

[35] The treatises on temperature and colour mixing were also included in Mayer (1775); Mayer's theory of the magnet was first published in Forbes (1972, III).

[36] Carsten Niebuhr was a talented student of Mayer's. He went as a surveyor with a disastrous Danish expedition, planned by Michaelis, to Arabia Felix, and he was the only one to survive. The party travelled from Denmark to the Middle East by ship and Niebuhr performed his lunar distance trials underway. He transmitted the results of the sea trials by letter to Mayer. For further reading on this fascinating expedition, see Hansen (1964). Niebuhr's reminiscences of his study period, recounted in various contributions to Von Zach's *Monatliche Korrespondenz*, form a valuable source of information on Mayer. The Niebuhr results are included in Mayer (1770, p. cxxvi).

A copy of the kil tables went also with Nevil Maskelyne overseas to Saint Helena in 1761. The object of this expedition was to observe the upcoming transit of Venus from that island. While underway, Maskelyne used the lunar tables successfully to experiment with the lunar distance method of longitude determination, and on the way home he taught the method to the ships' officers. Upon his return he prepared and published the *British Mariner's Guide*, a manual of the lunar distance method for seamen, with the kil lunar tables appended. Gael Morris had adjusted the epochs of the tables to the meridian of the Greenwich Observatory, whereas Mayer's were based on the meridian of the Observatory of Paris.[37]

Likewise, the Abbé Nicholas Louis de Lacaille had experimented with lunar distances in 1753 on the way home from his outpost at the Cape of Good Hope, though with less accurate lunar tables. Perhaps he made use of the tables of Lemonnier, which we will discuss in Chap. 6. Lacaille expounded a method to compute longitude from an observed lunar distance in the *Connoissance des Temps* for the year 1761, and Mayer's kil tables found their way into that volume, too.[38]

With the war drawing to a close, the Admiralty saw fit to submit Mayer's and Harrison's methods of longitude finding to a final test at sea. The necessary arrangements were made and the tests performed on a voyage to Barbados in the West-Indies and back, with – as might be expected with so much at stake – considerable quarrels, which need not be mentioned here.[39]

Eventually, in an illustrious meeting held on 9 February 1765, the Board of Longitude advised to the British Parliament that both John Harrison (now over 70) and Tobias Mayer (already deceased, in the same year as Bradley, Lord Anson, and Lacaille) should be rewarded for their outstanding contributions to the longitude problem. The former was to receive £10.000 or half the maximum reward, because although his chronometer had performed well within the limits stipulated for the maximum reward, it was not yet clear how to produce enough copies of it for a whole fleet; if Harrison could demonstrate how to lift this limitation, he would also qualify for the second half. Mayer's widow received £3000: her late husband's method was accurate enough to achieve the lower limit of 60 nautical miles set by the Longitude Act, but the necessary calculations were rather prolix, and the instrument that Mayer had proposed was considered less suitable than the newly developed sextant.

An unexpected reward of £300 went to Leonhard Euler for his alleged contribution to Mayer's lunar theory. Perhaps the British Parliament reached this decision after protests of Clairaut that not Mayer, but Euler and himself

---

[37] Maskelyne (1761, 1763); see also Howse (1989, Chaps. 3–5). Howse's remark (p. 42) that Maskelyne published the kil tables are easily disproved by Maskelyne's own comments in Maskelyne (1763, p. 123).

[38] Lacaille (1759).

[39] See the standard literature on this topic, e.g., Andrewes (1996), Howse (1980), Forbes (1975).

had conceived that theory.[40] If so, their argument can now be seen to be false, because, as we will see, Mayer's tables depended less on that theory than was apparent in those days.

At the same meeting of the Board in February 1765, Nevil Maskelyne, who was appointed to the office of Astronomer Royal the very day before (and henceforth he was an ex-officio member of the Board),[41] was charged with a project of his own invention: namely, the production of tables and computational tools to assist the mariner in finding his longitude by the lunar distance method. Maskelyne planned to produce two volumes of tables. One volume would contain an almanac with ephemeris data to appear annually, the other the permanent tables for refraction, parallax in altitude, tables of logarithms of trigonometric functions, and various tables necessary for the reduction of lunar distances.

The novelty in the almanac would be the inclusion of precomputed lunar distance tables showing the angular distances of the moon from the sun and selected stars for every 3 h of local time on the meridian of the Royal Greenwich Observatory. Lacaille had proposed such tabulations, and example tables covering a fortnight had been appended to his exposition of the method of lunar distances in the *Connoissance des Temps*[42]; they would take away most of the computational burden of the navigator wishing to find his longitude by lunar distance. Maskelyne's project materialized in *The Nautical Almanac and Astronomical Ephemeris* and *Tables Requisite to be Used with the Astronomical and Nautical Ephemeris*; both first appeared in 1767. The Nautical Almanac is still in production today, albeit under a revised name and with its contents adapted to more modern needs.

During the first decades of its existence, the lunar tables in the Almanac were computed from **rede**, Mayer's last manuscript tables, with (from 1777 on) several improvements and additions by Charles Mason (formerly an assistant of Bradley). Maskelyne prepared the **rede** tables for the press, and they were published with revised epochs, three years after he published Mayer's account of the lunar theory. Other authors adapted these tables for their own publications.[43] Mayer's manuscripts of these tables are now kept in the Royal Greenwich Observatory Archives.[44]

---

[40] For example, Forbes (1975, p. 124), Clairaut (1752b, 2nd ed., p. 102), quoted on p. 58 below. Clairaut also made a weak attempt to fix the Board's attention on his own lunar theory as a worthy alternative to Mayer's, in a letter directed to John Bevis (*Gentleman's Magazine*, May 1765, p. 208).

[41] Maskelyne succeeded Nathaniel Bliss, who had succeeded Bradley after his death in 1762.

[42] Cf. fn. 38.

[43] For example, Hell and Pilgram (1772), the second edition of Lalande (1764) (the first edition contained Mayer's kil tables; Lalande also applied them for the computation of the data in the *Connoissance des Temps*, the French almanac), and Steenstra (1771).

[44] RGO 4/125, see appendix of consulted manuscripts.

Mason improved Mayer's tables a second time, fitting them to almost 1,200 lunar observations of Bradley from the 1750s, and he added tables for eight equAtions (sic) that Mayer had included in his theory, but not in his tables. Mason's new tables were still in the multistepped format; they were used, with slight alterations and additions by Lalande, for the almanacs from 1789 to 1812. Mason's tables in turn were improved by Bürg, who added still more equAtions from Mayer's theory and used more than 3,200 lunar observations of Nevil Maskelyne to fit the coefficients. He too retained the multistepped format.[45] Mayer's *solar* tables, which were essentially Lacaille's of 1758 with slight adjustments, were used for the almanacs from 1767 to 1804.[46]

The lunar distance method was, for most navigators, the only way of longitude determination at sea until chronometers were produced in sufficient quantities and their prices dropped, in the second quart of the nineteenth century. And even then 'lunars' and chronometers co-existed: the former, if handled by a skilled observer, being the only way to check the rate of a chronometer during long voyages.

Although his tables were gradually replaced, two important aspects of Mayer's work persisted. Their multistepped format survived into the early years of the nineteenth century before it was abandoned. The principle of amassing data to fit parameters, on the other hand, is now firmly established in scientific practice.

---

[45] See Mason (1787). I have not looked into Mason's methods to improve the tables; anyone intending to do so should consult RGO 4/193, 2nd part, containing a draft of Maskelyne's instructions to Mason. Bürg's tables are contained in Delambre (1806). Some confusion exists about the period 1809–1812. In RGO (1821, pp. 78–79) (also quoted in Forbes (1965, p. 399)) Thomas Young, secretary of the Almanac, remarks that Mason's (probably revised) tables were used; Seidelmann (1992) is vague about this period and maintains that the almanacs since 1813 were based on Bürg's tables of Laplace's, not Mayer's, theory.

[46] Forbes and Wilson (1995, p. 61), Seidelmann (1992); Wilson argues that Mayer's adjustments to Lacaille's coefficients were no improvement (Wilson 1980, p. 188).

Chapter 4
A Manual to the Tables

In preparation of the technical chapters ahead, the reader is made familiar with the terminology of the *multistepped* and *single-stepped* formats of lunar tables. These concepts address the way in which the tables are put to use. They have nothing to do with the way by which their coefficients were fixed, and also one should not associate them with some kind of an iterative scheme. The term 'multistep' refers to the fact that the computation of a lunar position, by way of these tables, goes through several steps, whereas 'single-step' refers to a computational procedure of only one step. Thus, 'step' is the key concept to master in this chapter.

After an introduction to the tables and an explanation of the multistep format, an example computation is given. A grasp of the structure of the computation will be found helpful in Chap. 7. Section 4.3, on time and calendar conventions, is of help to anyone who intends to work with the tables, but it is not important for an understanding of the later chapters or of the format.

We will use the tables of the rede edition here, because these tables were more widely distributed and because they fit better than the kil version to the end result of Mayer's lunar theory, which we encounter in Chaps. 5 and 7. Both kil and rede were accompanied by an example calculation when published.[1] The last section of this chapter points out in what way the format of those two sets of tables differs.

4.1 On Tables

Several tables are needed for a single position calculation of the moon. Some of these tables represent a mean position or mean motion: these tables are entered with a date/time argument. Other tables represent an equation: these

---

[1] Mayer had warned that the example computation included with the 1753 kil tables was based on a slightly older version, which I could identify as zwin.

S.A. Wepster, *Between Theory and Observations*, Sources and Studies in the History of Mathematics and Physical Sciences, DOI 10.1007/978-1-4419-1314-2_4,
© Springer Science+Business Media, LLC 2010

tables are entered with an argument in angular measure. Since most of the
lunar equAtions depend on the position of the sun, the sun's position is needed
aswell and it is computed from tables en passant. Appendix A lists all the
mean motions and equAtions of several table versions including rede.

We recall that from a modern mathematical point of view, an equAtion is
a function; all equAtions of longitude are of the form $\sum_{k=1}^{n} c_k \sin(k\alpha)$ with
(as far as we are concerned) $n \leq 4$. The c_k are coefficients and α is called
the argument of the equAtion. Mayer's tables are effectively lists of function
argument and function value pairs: see Fig. 4.1 for examples. A table is used
to look up the value of the function for a specific value of its argument; this
value is then added to some other argument, which it serves to correct, or to
equAte in the old terminology, as will be shown in detail in Sect. 4.4.

The functions that the mean motion and epoch tables represent, are linear
functions, while the table for secular acceleration provides a quadratic term.
An example of the epoch tables showing the mean positions at the beginning
of the years 1780–1807 is contained in Fig. 4.1.

The following is a complete list of the 22 tables in rede necessary to compute the longitude of the moon. Because we will not consider the calculations
for the moon's parallax and latitude, which are somewhat similar to the longitude computation, there is no need to list the tables for their calculation
here. The solar tables are likewise omitted, although an example computation
of the sun's longitude is given further down.

- Mean motion of the moon's longitude, longitude of the apogee, and longitude of the ascending node for integer numbers of elapsed Julian years (i.e,
  years of 365 days, and a leap day for every fourth year; note that these
  are time intervals, not years in the Julian calendar).
- Mean epochs of longitude, apogee, and node for the beginning of the year
  (more precisely, for mean noon on the preceding 31 December) on the
  Greenwich meridian. These were listed for selected years in both the Julian
  and Gregorian calendars.[2]
- Mean motions of longitude, apogee, and node for months and days.
- Mean motions of longitude, apogee, and node for hours, minutes, and
  seconds.
- Secular acceleration, from 900 BC to AD 4300.
- Annual equAtion of the anomaly as a function of mean solar anomaly.
- Annual equAtion of the mean ascending node as a function of mean solar
  anomaly.
- Thirteen tables (numbered from I to XIII) for just as many equAtions, all
  with different arguments. These include the annual equAtion of longitude,
  equAtion of centre, evection, and variation. Further details are provided in
  the example in Sect. 4.4.

---

[2] The Julian calendar table listed epochs of every century from 600 BC to AD 1400, every
20 years from 1460 to 1700, and every year from 1701 to 1752 (when Great Britain switched
to the Gregorian calendar); the Gregorian calendar table listed epochs of every 20 years
from 1600 to 1700, and every year from 1701 to 1807.

Epochæ mediorum motuum LUNÆ,
Tempore medio currente sub æridiano Observatorii GÆNOVICENSIS.

Stylo Gregoriano	Longit. med. ☽ (s o ′ ″)	Long. Apóg. ☽ (s o ′ ″)	Longit. ☊ (s o ′ ″)
Anni post C.N. { 1780	7. 5.19.46	10.11.44.37	2. 0. 2.10
1781	11.14.42.51	11.22.24.27	1.10.42.27
1782	3.24. 5.57	1. 3. 4.18	0.21.22.44
1783	8. 3.29. 2	2.13.44. 8	0. 2. 3. 1
1784	0.26. 2.43	3.24.30.40	11.12.40. 7
1785	5. 5.25.48	5. 5.10.30	10.23.20.24
1786	9.14.48.53	6.15.50.21	10. 4. 0.41
1787	1.24.11.59	7.26.30.11	9.14.40.58
1788	6.16.45.39	9. 7.16.43	8.25.18. 4
1789	10.26. 8.45	10.17.56.33	8. 5.58.21
1790	3. 5.31.50	11.28.36.24	7.16.38.38
1791	7.14.54.55	1. 9.16.14	6.27.18.55
1792	0. 7.28.36	2.20. 1.46	6. 7.56. 1
1793	4.16.51.41	4. 0.42.36	5.18.36.18
1794	8.26.14.47	5.11.22.27	4.29.16.35
1795	1. 5.37.52	6.22. 2.17	4. 9.56.52
1796	5.28.11.32	8. 2.48.49	3.20.33.58
1797	10. 7.34.38	9.13.28.39	3. 1.14.15
1798	2.16.57.43	10.24. 8.30	2.11.54.32
1799	6.26.20.48	0. 4.48.20	1.22.34.49
1800	11. 5.43.54	1.15.28.10	1. 3.14. 6
1801	3.15. 6.59	2.26. 8. 0	0.13.54.23
1802	7.24.30. 5	4. 6.47.51	11.24.34.40
1803	0. 3.53.10	5.17.27.41	11. 5.14.57
1804	4.26.26.51	6.28.14.13	10.15.52. 3
1805	9. 5.49.56	8. 8.54. 3	9.26.32.20
1806	1.15.13. 1	9.19.33.54	9. 7.12.37
1807	5.24.36. 7	11. 0.13.44	8.17.52.54

I. Pro longitudine Lunæ.
Argum. I. Anomalia media Solis.

S.	5 + (′ ″)	4 + (′ ″)	3 + (′ ″)	2 + (′ ″)	1 + (′ ″)	0 + (′ ″)	S.
30	5.42	9.49	11.16	9.42	5.35	0. 0	0
29	5.31	9.43	11.16	9.48	5.45	0.12	1
28	5.21	9.37	11.16	9.54	5.54	0.23	2
27	5.10	9.31	11.15	9.59	6. 4	0.35	3
26	4.59	9.24	11.14	10. 5	6.14	0.47	4
25	4.49	9.18	11.14	10.10	6.24	0.58	5
24	4.38	9.11	11.12	10.14	6.33	1.10	6
23	4.27	9. 4	11.11	10.19	6.43	1.21	7
22	4.16	8.57	11. 9	10.24	6.52	1.33	8
21	4. 5	8.49	11. 7	10.28	7. 1	1.45	9
20	3.54	8.42	11. 5	10.33	7.10	1.56	10
19	3.43	8.34	11. 3	10.37	7.19	2. 7	11
18	3.31	8.26	11. 0	10.44	7.28	2.19	12
17	3.20	8.18	10.58	10.48	7.37	2.30	13
16	3. 8	8.10	10.55	10.51	7.46	2.42	14
15	2.57	8. 2	10.52	10.54	7.54	2.53	15
14	2.45	7.54	10.49	10.57	8. 2	3. 4	16
13	2.33	7.45	10.45	10.59	8.10	3.15	17
12	2.21	7.36	10.42	11. 2	8.18	3.26	18
11	2.10	7.27	10.38	11. 4	8.26	3.38	19
10	1.59	7.18	10.34	11. 8	8.34	3.49	20
9	1.47	7. 9	10.30	11.10	8.41	3.59	21
8	1.35	7. 0	10.25	11.11	8.49	4.10	22
7	1.23	6.51	10.21	11.13	8.56	4.21	23
6	1.11	6.41	10.16	11.14	9. 3	4.32	24
5	1. 0	6.32	10.11	11.15	9.10	4.43	25
4	0.48	6.22	10. 6	11.15	9.16	4.53	26
3	0.36	6.12	10. 0	11.16	9.23	5. 4	27
2	0.24	6. 2	9.55	11.16	9.30	5.14	28
1	0.12	5.52	9.52	11.16	9.36	5.24	29
0	0. 0	5.42	9.49	11.16	9.42	5.35	30
S.	6	7	8	9	10	11	S.

Fig. 4.1 Example tables from Mayer (1770). *Top:* p. XLII, epochs. The errata mentioned that the node positions of 1800 and later were 1′ too small, but an identical error in the 1792 apogee went unnoticed. *Bottom:* p. LIII, the annual equation, with mean solar anomaly as argument

- Two more tables, to adjust the frame of reference rather than the position of the moon. The first is for the so-called reduction, which converts the moon's longitude in its orbital plane to the plane of the ecliptic. The second is the equAtion of the equinoxes. It shifts the origin of the longitude scale from the mean equinox of date to the true equinox of date, thus correcting for nutation. There is no need for the reduction from the mean equinox of epoch to the mean equinox of date, because the mean motion tables already include precession. Earlier tables of Mayer's such as kil had no nutation correction.

4.2 Single-Step and Multistep Procedures

Modern lunar theories are usually formulated in such a way that all the equAtions (i.e., all their periodic terms) are computed directly from the mean motion arguments.[3] This is what we call here the single-stepped procedure. Alternatively, one may imagine that the arguments are equAted, i.e., adjusted, after each equAtion is computed and before the next one, or, to put it differently, that the position of the moon is improved step by step from the mean towards the true place, adding one equAtion at a time. Each next equAtion is then computed with updated arguments. Newton's lunar theory of which we shall speak in Chap. 6 is such a theory. Indeed, when one starts out from the differential equations of motion and proceeds in true eighteenth century style, then a single-stepped theory will be the result.

The lunar tables of Mayer may be distinguished according to the procedure that should be followed when using them. His older tables up to early in 1752 are of the 'modern' single-stepped form: all their equAtions should be computed with the same mean motion arguments. The lunar tables of his colleagues Euler, Clairaut, and d'Alembert were of this kind, too. Even Mayer's treatise *Theoria Lunae*, written in 1755, initially developed a single-stepped theory.

But from the spring of 1752 onwards, Mayer's tables occupy a position in between the single-stepped form and the step-by-step form of Newton's theory. Leaving it to Chap. 6 to demonstrate the link between Newton's and Mayer's theories, we may here presuppose that Mayer reordered Newton's equAtions and lumped several of them together in a smaller number of 'steps'. Such is the idea of what I have called the multistep form of his tables.

Among those multistepped tables, different varieties may be distinguished once more. The most important variants are the rede tables, which we will employ in our example calculation, and the kil version. They differ by the

---

[3] Examples include almost all dynamical theories such as ELP2000 or Laplace, described in Chapront-Touzé and Chapront (1983) and Laplace (1802), respectively.

way in which the equAtions are distributed over the steps. Incidentally, this also has consequences for the magnitudes of the equAtions's coefficients, as will be explored further in Chaps. 6 and 7.

We will now present the steps of the rede tables. In the last section of the chapter they will be compared to the slightly different steps of the kil version.

First step: Compute the mean motion arguments, ten so-called 'minor equAtions,' and two annual equAtions of apogee and node.

Second step: Update (i.e., adjust) the arguments with the ten minor and the annual equAtions, then use the updated anomaly as argument for the equAtion of centre.

Third step: Update the lunar longitude with the equAtion of centre. Use the thus updated longitude to find the variation.

Fourth step: Update the longitude with the result, and find out the value of the XIII equAtion, which depends on longitude, node, and anomaly.

Fifth step: Update again, and find the reduction and the equAtion of the equinoxes before finally arriving at the true longitude of the moon.

The essence of the multistep procedure is that the arguments in each step differ from the arguments in the previous one, the difference being brought about by the application of one or more equAtions. It is, however, not essential whether we decide that a new step begins just before or just after the arguments are updated. The example computation to be discussed below follows Maskelyne's rules, where the updating of individual arguments is delayed until they are needed. This streamlines the work flow without affecting the end result.[4]

4.3 Calendar Conventions

The time to be used with the epoch and mean motion tables of rede is mean solar time on the Greenwich meridian. Mayer had given his epochs for mean time on the meridian of the Royal Observatory in Paris, but Nevil Maskelyne adjusted the values to fit his own workplace at the Royal Greenwich Observatory. The astronomical day customarily started at 12 noon, half a day later than the civil day, so that hours from 0 to 12 refer to an event post meridiem, while hours from 12 to 24 likewise refer to an event ante meridiem on the next civil calendar date.

---

[4] Maskelyne included both Mayer's rules and his own modifications in the explanatory chapter of Mayer (1770). The earlier tables in Mayer (1753b) had no explanation of procedure, only an example computation. A sketch of the procedure is also in Forbes and Wilson (1995, pp. 64–65). Maskelyne also published instructions and an example in Maskelyne (1763, pp. 28–37) for the kil tables contained in that work.

Maskelyne's time scale is roughly equivalent to the definition of Greenwich Mean Time (GMT) prior to 1925. From 1925 on, day start was redefined at midnight, and soon afterwards Universal Time (UT) was introduced, satisfying the relation UT − GMT = 12 hours. The GMT and UT time scales are both coupled to the rotation of the earth. In the middle of the nineteenth century, Adams and Delauney confirmed earlier suspicions that the earth's rotation is not uniform. Accuracy demands have since then led to an increasing call for a uniform time scale on which to base orbit computations. This has resulted in the 1950s in the design of Terrestrial Time (TT), to be used for computations of apparent geocentric ephemerides.

The non-uniformity of the earth's rotation is responsible for the difference $\Delta T = \text{TT} - \text{UT}$. The earth's rotation, and hence the value of ΔT, changes slowly because of tidal friction and various other only partly understood processes. The ΔT-value for 2002 was +64s; extending the concept to historical times yields a value of +15s in 1755. This difference matters when comparing historical lunar observations and computations to modern computations, because in 1 min of time the moon moves about $30''$ of arc with respect to the stars. The discrepancies between a uniform timescale (TT) and a non-uniform one (UT) show most prominently in the position of the fastest moving celestial body, the moon.[5]

Leap days were intercalated in Mayer's tables as the first day in the year (Jan. 0, so to say), instead of between 28 February and 1 March. A user of the tables had to subtract one day if the date for which he made a computation happened to be in the months January and February of a leap year.

4.4 An Example Computation

Every calculation of a lunar position starts with looking up five mean motions (two for the sun and three for the moon), and sometimes a secular acceleration term. From these mean motions one computes the four basic arguments: ω, the lunar longitude minus the solar longitude; p, the lunar longitude minus the longitude of lunar apogee; δ, the lunar longitude minus the longitude of its node; and ς, the distance of the mean sun from the solar apogee. The arguments of all equAtions are simple integer linear combinations of these four.[6] Next in the procedure, the lunar anomaly and node are corrected by their annual equAtions, and the lunar longitude by an aggregate of about ten so-called 'minor equAtions'. The arguments of these equAtions are neatly listed together with the values of the equAtions answering to those arguments, and the sum of the values is taken. The next steps in the calculation take care

---

[5] The concepts of time are covered in Seidelmann (1992, Chap. 2); refer to p. 11 for tidal friction and secular acceleration.

[6] Some later versions of Mayer's tables have one exception in the form of an equAtion depending on the longitude of the ascending node of the lunar orbit; see fn. 18 on p. 33.

of the remaining equAtions, principally the equAtion of centre, variation, and reduction, in a similar way. Some table versions include the evection among the 'minor equAtions', others treat evection separately in a later step, as will be discussed more fully in the last section of the chapter.

The following example shows in detail how to compute the ecliptic longitude of the moon for 2002 November 15, 08:00:00 UT from the rede tables. The sun's position is also computed, because several arguments of the lunar tables depend on its position. The description of the calculation is a paraphrase of Maskelyne's translation of Mayer's instructions. For ease of reference to the cited source, the numbering in square brackets of the following paragraphs is Maskelyne's.[7] His numbering has nothing to do with the division in steps of the multistepped procedure.

4.4.1 To Find the Sun's Longitude

The solar tables that came with rede are Mayer's but adjusted to the time of the Greenwich meridian; Mayer's tables in turn are a slight modification from Lacaille's solar tables.[8]

The tables use sexagesimal notation of degrees, minutes, and seconds. Along the ecliptic, however, arcs were further subdivided into *signs*, obeying an age-old practice, with 1 sign = 1^s = $30°$. Full circles of 12^s are silently discarded from calculations, or, phrased differently, angles along the ecliptic and equator are taken modulo 12^s.

The following explanation of the computation of the sun's true longitude and mean anomaly can be quickly scanned over by a reader interested in the workings of the lunar tables only. But since computations for the sun are so much easier, this part could be of help when difficulties are experienced with the lunar part of the calculations.

Display 4.1 shows the completed calculation for the sun's true longitude. The numbers on the left side of the display correspond to the numbered instructions of the explanation.

[1] *(Epoch)* Make one column for the sun's longitude, and one for its apogee and anomaly. Look up in the tables the mean longitude of the sun and of its apogee for the given year. (If the year is outside the range of the table, then take the values for a convenient earlier year, and look up in the table for single (Julian) years the mean motions for the remaining years.)

[2] *(Mean motion)* Take from the tables the mean motions of the longitude and apogee for the month, day, and time of day, and write them down under the mean longitude and apogee just found.

---

[7] Cf. Mayer (1770), pp. 101–102 for the sun, pp. 121–127 for the moon.

[8] Wilson (1980, pp. 184–188).

Display 4.1 The completed calculation of the solar longitude

		Longitude	Apog./anom.
[1]	epoch 1802	9^s 9° 25′ 56.6″ t	3^s 9° 34′ 46″ t
[1]	200 years	0^s 1° 32′ 46.0″ t	0^s 3° 40′ 0″ t
[2]	Nov 13	10^s 12° 27′ 0.9″ t	57″ t
[2]	20^h	49′ 16.9″ t	
[3]	mean longitude	7^s 24° 15′ 0.4″	3^s 13° 15′ 43″
[4]	mean anomaly ς		4^s 10° 59′ 17″
[5]	eqn. of centre	−1° 28′ 24.6″ t	
[6]	small eqns	−5.1″ t	
[7]	true longitude ☉	7^s 22° 46′ 30.7″	

The letter 't' indicates values that have been looked up in a table.

[3] *(Sun's mean position)* Add the numbers in the columns to find the mean longitude of the sun and of its apogee for the given date and time (remember to work modulo 12^s).

[4] *(Mean anomaly)* Subtract the longitude of apogee from the longitude of sun, and write the result under the apogee column. This is the mean anomaly of the sun.

[5] *(EquAtion of centre)* With this mean anomaly as argument, enter the table of the equAtion of the centre, look up the value of this equAtion and place it under the sun's mean longitude.

[6] *(Planetary perturbations)* From appropriate columns in the solar mean motion tables, find three small equAtions due to the perturbing effects of the moon, Jupiter, and Venus on the earth, and one for the equAtion of the equinoxes. For brevity and clarity, I omit the details of these small effects.

Note that at this stage we encounter two equAtions depending on the moon, whose exact position is not known without recourse to the *lunar* tables. But since the absolute values of these equAtions are at most 8″ and 18″, respectively, it suffices to take the position of the mean moon here.

[7] *(Sun's true longitude)* Add the equAtions of [6] to the mean longitude of the sun to arrive at its true longitude.

Our specific example of 2002 November 15, 08:00 UT runs well outside the range that Nevil Maskelyne had catered for in his epoch table, which went no further than 1807. Therefore, we take in [1] the epoch year 1802 and add 200 Julian years from the table for single years. Then we subtract one day from the given date, because 1900 is a leap year in the Julian calendar, but not in the Gregorian calendar. In [2] we also subtract 12 h from the given time, because the astronomical day started conventionally 12 h after midnight. Thus, we enter the tables with November 13, 20:00 GMT. The computation, when properly carried out, should look like Display 4.1.

Display 4.2 Calculation of mean arguments of the moon

		Longitude	Apogee	Node
[2]	sec. accel	$1'\,21''$ t		
[2]	epoch 1802	$7^{\mathrm{s}}\,24°\,30'\ \ 5''$ t	$4^{\mathrm{s}}\ \,6°\,47'\,51''$ t	$11^{\mathrm{s}}\,24°\,35'\,40''$ t
[2]	200 years	$8^{\mathrm{s}}\,15°\,47'\,10''$ t	$7^{\mathrm{s}}\ \,8°\,22'\,30''$ t	$8^{\mathrm{s}}\,28°\,22'\,30''$ t
[2]	Nov 13	$7^{\mathrm{s}}\ \,6°\,55'\ \ 4''$ t	$1^{\mathrm{s}}\ \,5°\,18'\,59''$ t	$16°\,47'\,12''$ t
[2]	20^{h}	$10°\,58'\,49''$ t	$5'\,34''$ t	$2'\,39''$ t
[3]				$9^{\mathrm{s}}\,15°\,12'\,21''$
[3]	mean long ☽	$11^{\mathrm{s}}\,28°\,12'\,29''$	$0^{\mathrm{s}}\,20°\,34'\,54''$	$2^{\mathrm{s}}\ \,9°\,23'\,19''\,☊$
[4]	mean anom p		$11^{\mathrm{s}}\ \,7°\,37'\,35''$	
[4]	elongation ω	$4^{\mathrm{s}}\ \,5°\,25'\,58''$		

The letter 't' indicates values that have been looked up in a table.

4.4.2 To Find the Moon's Longitude

Now we come to a calculation of the moon's position, a slightly more complicated affair. We recall that we were to compute the lunar true longitude on
2002 November 15, 08:00 UT, and that we enter the tables with November
13, 20:00 GMT because of the calendar conventions. In the following computation, one readily recognizes the equation of centre, evection, variation,
and the annual equation. Displays 4.2–4.4 show several stadia of the computation. As before, the numbers in square brackets correspond to Maskelyne's
explanations; they do not match the sequence of steps outlined above. The
completed calculation is shown in Display 4.4.

[1] *(Sun's true longitude and mean anomaly)* Find the true longitude and
mean anomaly of the sun as already explained in the previous section (this
step is not represented in Display 4.2, which concerns the lunar arguments
only).

[2] *(Epoch and mean motion)* The first part of the computation is analogous to finding the sun's mean position. Look up the desired year in the
moon's epoch tables and take out the mean longitudes of the moon, her
apogee, and ascending node; write these numbers next to each other in three
columns. Place under them the mean motions in longitude, apogee, and node
taken from the mean motion tables for the month, day, and time of day. Pre-
fix the secular acceleration (given in a separate table) to the column of the
moon's longitude.[9]

[3] *(Mean position)* Add the numbers in the first column to find the moon's
mean longitude, and add the numbers in the second column to find the mean
longitude of apogee. Then find the mean longitude of the ascending node

---

[9] Strictly, what is here called the 'longitude' of the moon is an angle in the lunar orbit,
not along the ecliptic. Thus it is not one of the standard ecliptic coordinates until the
reduction in [8] below has been applied.

Display 4.3 The minor equations in the computation of lunar longitude

Table	Argument		Value
I	ς	$4^{\mathrm{s}}\ 10°\ 59'$	$+\ 8'\ 34''$
II	$2\omega + \varsigma$	$0^{\mathrm{s}}\ 21°\ 51'$	$-\ 0'\ 20''$
III	$2\omega - \varsigma$	$3^{\mathrm{s}}\ 29°\ 53'$	$-\ 1'\ 0''$
IV	$2\omega + p$	$7^{\mathrm{s}}\ 18°\ 30'$	$-\ 0'\ 40''$
V	$2\omega - p$	$9^{\mathrm{s}}\ \ 3°\ 14'$	$+1°20'\ 21''$
VI	$2\omega - p + \varsigma$	$1^{\mathrm{s}}\ 14°\ 13'$	$+\ 1'\ 30''$
VII	$2\omega - p - \varsigma$	$4^{\mathrm{s}}\ 22°\ 15'$	$+\ 0'\ 30''$
VIII	$p - \varsigma$	$6^{\mathrm{s}}\ 26°\ 39'$	$-\ 0'\ 15''$
IX	$\Omega - \odot$	$6^{\mathrm{s}}\ 16°\ 37'$	$+\ 0'\ 32''$
X	$\omega - p$	$4^{\mathrm{s}}\ 27°\ 48'$	$+\ 1'\ 2''$
[4]			$+1°30'\ 14''$

All values in the right hand column have been looked
up in tables.

Ω by *subtracting* the mean motions in the third column from the epoch of
the node. These must be subtracted because the node moves retrograde or
backwards.[10]

[4] *(Anomaly and elongation, and ten minor equations)* Subtract the longitude of apogee from the longitude of the moon. The difference is the mean
anomaly of the moon; write this number under the apogee column. Also subtract the true longitude of the sun from the mean longitude of the moon, the
result is the elongation of the moon.[11] Display 4.2 shows the calculation for
the three basic mean arguments: mean lunar longitude, mean anomaly, and
mean longitude of the ascending node. With the mean motion values thus
found, form ten different arguments for ten different tables of what Mayer
called the *minor equations*. Nine of these ten arguments are integer linear
combinations of the three arguments that we have computed already (see
Display 4.1): the mean solar anomaly ς, elongation ω, and mean lunar anomaly p; an additional argument, $\Omega - \odot$ (see Displays 4.1 and 4.2), expresses the
orientation of the lunar line of nodes with respect to the direction of the sun.
Look up the equation values in the tables for each of these ten equations,
and write them down as in Display 4.3. Add the equation values and write
their sum at the bottom.

The Roman numerals correspond to the Roman numbering of the tables.
The I equation is the annual equation of longitude; the V is the evection.

[5] *(Annual equations of anomaly and node, and equation of centre)* With
argument ς, look up the annual equations of the moon's mean anomaly and
ascending node. Next, we enter the second step of the multistep procedure,

---

[10] To make the calculations easier and more uniform, some later tables (e.g., Lalande
(1764); Hell and Pilgram (1772)) tabulated the complement of the longitude of the ascending node, which increases in time just like longitude and mean apogee.

[11] Actually, the term *elongation* is ambiguous, since it could mean either the difference in
longitude of a body and the sun, or the angular separation of a body from the sun (see
also (Meeus 1998, p. 253)). In this text, elongation will be the difference of the ecliptic
longitude of the true sun and the longitude of the mean moon in its orbit.

Display 4.4 The completed calculation of lunar longitude

		Longitude	Apogee	Node
[2]	sec. accel	$1'\,21''$ t		
[2]	epoch 1802	$7^{\rm s}\,24°\,30'\,5''$ t	$4^{\rm s}\,6°\,47'\,51''$ t	$11^{\rm s}\,24°\,35'\,40''$ t
[2]	200 years	$8^{\rm s}\,15°\,47'\,10''$ t	$7^{\rm s}\,8°\,22'\,30''$ t	$8^{\rm s}\,28°\,22'\,30''$ t
[2]	Nov 13	$7^{\rm s}\,6°\,55'\,4''$ t	$1^{\rm s}\,5°\,18'\,59''$ t	$16°\,47'\,12''$ t
[2]	$20^{\rm h}$	$10°\,58'\,49''$ t	$5'\,34''$ t	$2'\,39''$ t
[3]				$9^{\rm s}\,15°\,12'\,21''$
[3]	mean long	$11^{\rm s}\,28°\,12'\,29''$	$20°\,34'\,54''$	$2^{\rm s}\,9°\,23'\,19''$
[4]	mean anom p		$11^{\rm s}\,7°\,37'\,35''$	
[4]	elongation ω	$4^{\rm s}\,5°\,25'\,58''$		
[5]	annual eqns.		$17'\,36''$ t	$6'\,42''$ t
[5]	minor eqns.	$1°\,30'\,14''$	$1°\,30'\,14''$	
[5]	cor. anom $\tilde{p}$		$11^{\rm s}\,9°\,25'\,25''$	
[5]	eqn. centre	$2°\,5'\,0''$ t		
[6,7]	cor. elong $\tilde{\omega}$	$4^{\rm s}\,9°\,1'\,12''$		$2^{\rm s}\,9°\,30'\,01''$ cor. Ω
[6]	variation	$-36'\,21''$ t		
[6]	cor. long	$0^{\rm s}\,1°\,11'\,22''$		$0^{\rm s}\,1°\,11'\,22''$
[7]	δ			$9^{\rm s}\,21°\,41'\,21''$
[7]	$2\delta - \tilde{p}$			$8^{\rm s}\,3°\,57'\,17''$
[7]	eqn. XIII	$-1'\,15''$ t		$-1'\,15''$ t
[8]	reduction	$4'\,36''$ t		$9^{\rm s}\,21°\,40'\,6''\;\tilde{\delta}$
[9]	eqn. equinox	$-17''$ t		
[10]	eclipt. long	$0^{\rm s}\,1°\,14'\,26''$		

The letter 't' indicates values that have been looked up in a table.

because we will now update some of the arguments with the equAtions found thus far. Add the annual equAtion of anomaly and the sum of the ten minor equAtions just found to the moon's mean anomaly. This will yield the corrected anomaly of the moon denoted by $\tilde{p}$. Use this as the argument to look up the equAtion of centre in table XI.

[6] *(Variation)* Add the sum of the ten minor equAtions and the equAtion of centre to the elongation ω, to get the corrected elongation $\tilde{\omega}$. Here begins the third step of the multistepped procedure. With $\tilde{\omega}$ look up the equAtion of the variation in table XII. Add this variation, the equAtion of centre, and the ten minor equAtions to the mean longitude to obtain the corrected longitude.

[7] *(EquAtions to the node)* Now we come to the fourth step of the multistepped procedure. Correct the mean longitude of the ascending node by its annual equAtion found in [5]. Subtract the result from the moon's corrected longitude, to get the equAted distance of the moon from the node δ. Compute $2\delta - \tilde{p}$, this is the argument of the equAtion listed in the XIII table. Write the value of this equAtion under the moon's corrected longitude, and also under the node column.

[8] *(Reduction to the ecliptic)* Entering the fifth and final step of the multistepped procedure, add the XIII equAtion to the equAted distance from the node to get $\tilde{\delta}$, with which look up the equAtion of the reduction to the ecliptic in the XIV table. Write the result under the longitude column.

[9] *(EquAtion of the equinoxes)* With the mean longitude of the node ☊ as argument, look up the equAtion of the equinoxes in the XV table. Write the result under the value obtained in the previous step.

[10] *(True longitude of the moon)* Add the last three equAtions to the corrected longitude of the moon to get its true longitude in the ecliptic. This completes the calculation.

4.5 A Note on Accuracy

We found that the tables predict the longitude of the moon in the ecliptic on 2002 November 15, 08:00 UT to be $1°14'26''$, and the sun's longitude $232°46'30.7''$. A computation using modern software[12] yields $1°10'23''$ for the longitude of the moon, and $232°46'38''$ for the sun's longitude. The difference is $4'3''$ for the moon, and about $7''$ for the sun.

It is well known from dynamical systems theory that every lunar theory will eventually diverge from the real position of the moon. In this example computation for a date in 2002, we see that the prediction according to the eighteenth-century lunar theory differs by $4'3''$ from the modern prediction. That is almost $\frac{1}{7}$ of the moon's diameter, and comparable to the thickness of one or two coins seen from a distance of 10 m. Such a perhaps seemingly small divergence would make Mayer's lunar theory useless as the basis for the lunar distance method of finding geographical longitudes today – even if we would be content with the same margins that were acceptable in 1760. The result is also significantly worse than the accuracy of about $30''$ to $1'$ which Mayer aimed at.

Would the result of this single example be typical for the error values? To answer this and similar questions, I made a computer program to simulate some of the different lunar tables and theories of Mayer's. The program was used to generate daily predictions (according to these same rede tables) a few months before and after 2002 November 15, and to compare them to the modern predictions. The differences showed an almost monthly recurring pattern ranging from about $-1\frac{1}{2}'$ to $+4\frac{1}{2}'$. The result for the example above happens to be near an extreme.

Most of the error is brought about by the mean motions, which are slightly off the correct values, leading to a slowly increasing error in mean longitude, apogee, and node. Particularly the accumulated discrepancy in mean apogee (about $20'$) leads to a periodically varying error in lunar longitude, as is fully explained in Appendix B, page 218. In modern times, Mayer's rede tables generally perform better than our single example calculation suggested – although not as good as in their own times.

---

[12] JPL HORIZONS 3.12 Giorgini et al. (1996).

Display 4.5 The sequence of equAtions in kil tables of 1753 (left) and **rede** tables of 1770 (right); only the arguments of the equAtions listed

	kil	rede
I	ς	ς
II	$2\omega + \varsigma$	$2\omega + \varsigma$
III	$2\omega - \varsigma$	$2\omega - \varsigma$
IV	$2\omega - p + \varsigma$	$2\omega + p$
V	$2\omega - p - \varsigma$	$2\omega - p$
VI	$2\omega + p$	$2\omega - p + \varsigma$
VII	$2\delta - p$	$2\omega - p - \varsigma$
VIII	$p - \varsigma$	$p - \varsigma$
IX	$2\delta - 2\omega$	$2\delta - 2\omega$
X	$\omega - p$	$\omega - p$
XI	p	p
XII	$2\omega - p$	ω
XIII	ω	$2\delta - p$
XIV	2δ	2δ

The same procedure repeated for a time span in the 1750s shows only incidentally errors in excess of $1'$; table 6.3 on page 116 lists a standard deviation of $30''$ for these tables. We may draw the conclusion that Mayer's tables fulfilled his proclaimed expectations during his own era, but that their accuracy gets diluted over time through the drift in the mean motions, in particular the mean anomaly.

The prediction error that we found for the sun's position was about $7''$ (or 0.2% of its diameter), which is much more accurate than that for the moon. This need not surprise us, because the motion of the sun is much more regular and much easier to predict than the lunar motion.

4.6 The Wanderings of Evection

The kil tables of 1753 and the **rede** tables of 1770 use the same multistep procedure of computation, with only a seemingly small change in the sequence of the equAtions, apart from differences in the coefficients.

The change is illustrated in Display 4.5, which lists the arguments of the cquAtions in their kil and **rede** sequences, with steps separated by horizontal lines (mean motions are not shown). Since the reordering of equAtions within a step is inconsequential by our definition of a step, the noteworthy changes from kil to **rede** concern the positions of the arguments $2\omega - p$ (the argument of evection) and $2\delta - p$. These both move into another step. It can be deduced from internal evidence that the move of the $2\delta - p$ equation took effect before March 1754, and that evection relocated in about November 1754.

Why would Mayer have made these changes? In Chap. 6 it will be made apparent that kil's combined equAtion of centre and evection in its second step was a direct inheritance from Newton's lunar theory. For reasons that will also be made clear there, Mayer might possibly have had uneasy feelings about this when Bradley asked him to submit the theoretical basis of his tables. Therefore, the breaking up of the combination may have been appealing to Mayer.

But another reason can be given, one that is more in accord with Mayer's own words: the change in the order of equAtions leads to more elegant operations for the table user.[13] In the new situation, argument $2\omega - p$ takes its place next to $2\omega + p$, nicely preceded by the two arguments $2\omega \pm \varsigma$ and nicely followed by the pair $2\omega - p \pm \varsigma$. This arrangement is easier for the human mind to work with, especially during bulk computations for many different dates. But a similar argument for the move of $2\delta - p$ is less convincing.

These changes are almost trivial for the table user, but they are far from trivial for the table maker. They have consequences for the magnitudes of the equAtion coefficients: not only of those equAtions that are moved around, but also others as well. The technique to quantify the consequences is the subject matter of Chap. 7; let it suffice here to provide some intuition of what is at hand.

Evection is an equAtion with an extreme value of $\approx 1°21'$, only second in magnitude to the equAtion of centre with maximum $\approx 6°32'$, and followed by number three, variation, which may reach $\approx 43'$. The change that we are considering is whether evection is applied before or after the equAtion of centre; in particular, whether or not evection affects the argument p. If, regardless of the change, the equAtions would be maintained with the same coefficients, then the resulting value of the equAtion of centre might differ by as much as $8'$, which is clearly devastating if the required accuracy of the end result is $\frac{1}{2}'$. To find the correct values of the coefficients, one needs to consider also combined terms with arguments such as $(2\omega - p) - p = 2(\omega - p)$, which affects a term of equation X, and $(2\omega - p) + p = 2\omega$, which affects the main term of the variation. Both these terms change by about $4'$, which is so much that in general equAtion X reverses sign. The equation with argument $2\delta - p$ is much easier to move around because its coefficient is much smaller.

It is now time first to turn to Mayer's account of his lunar theory; thereafter we have the full and right perspective to tackle the ins and outs of the multistepped procedure.

---

[13] Mayer's words on this subject are 'But because the form of these [i.e., the tables in the kil format] is less appropriate, and [because] the calculation on behalf of the many equAtions that have thence to be taken out causes much trouble, I do not want to present them, and therefore I have sent only those that are easier to use. (*'Sed cum forma his sit minus commoda, et calculus propter plures aequationes inde excerpendas plus negotii facescat, nolui eas praeferre, atque adeo transmisi istas, quae usu faciliores sunt.'*) Mayer to Michaelis, 14 April 1755, reproduced in Mayer (1770, p. 43). This formed part of Mayer's response to inquiries on behalf of the Astronomer Royal James Bradley about tables that Mayer had *not* submitted for the Longitude Prize.

Chapter 5
Theoria Lunae

The subject of this chapter is *Theoria Lunae iuxta Systema Newtonianum*, a booklet published in London in 1767 by order of the Commissioners of Longitude. Its author was Tobias Mayer, and the text had been prepared for the press by Nevil Maskelyne, the Astronomer Royal.[1] We will undertake a study of this theory of Mayer's (*Theoria Lunae* for short) as well as the circumstances under which it was conceived, in order to re-evaluate, in the final chapter, its place in Mayer's work, particularly its relation to his lunar tables. As usual, our main focus will be directed towards the lunar longitude, taking latitude and parallax into account only where necessary.

5.1 Background

Theoria Lunae has been regarded, to varying degrees, as essentially Euler's lunar theory adjusted to observations.[2] Indeed there are some similarities between the two, at least when Euler's theory is taken to mean his *Theoria Motus Lunae* of 1753.[3] A comparison of the two theories, however, shows

---

[1] Mayer (1767). Mayer's manuscript is in RGO 4/108 (see appendix of consulted manuscripts).

[2] Examples include: Dictionary of Scientific Biography, Euler lemma, p. 481; (Lalande, 1764, 2nd ed., Vol. II, p. 224); Moulton (1902, p. 364); Pannekoek (1951, p. 251); Sadler (1977, p. 8); Cook (1998, p. 398). More circumspect are the viewpoints expressed in Brown (1896, p. 246), Forbes and Wilson (1995), and (less explicit) Linton (2004, p. 304). Waters makes the following curious remark: '...only in the 1760s did it (i.e., the lunar distance method) become feasible, when it was based on Newton's theory of the Moon, published in Gregory's *Astronomiae Physicae* of 1702' (Waters 1990, fn. 34 on p. 202). This is unlikely to be based on research, yet in Chap. 6 I argue that there is nevertheless quite some truth in his statement.

[3] Euler (1753). Several other lunar theories of Euler exist as well, including the tables in Euler (1746) and Euler (1772) in which he employed a rotating rectangular frame of coordinates.

S.A. Wepster, *Between Theory and Observations*, Sources and Studies in the History of Mathematics and Physical Sciences, DOI 10.1007/978-1-4419-1314-2_5, © Springer Science+Business Media, LLC 2010

that the similarity goes only as far as the construction of the differential equations of the moon's motion and the application of trigonometric series to find a solution.[4] The manner in which the solutions are reached is quite different in the two theories.

There are also specific points where Clairaut's essay *Théorie de la Lune*[5] may have been a source of inspiration to Mayer. Clairaut's theory won the prize contest of the Academy of Saint Petersburg of 1750, which posed to the participants the problem whether the inequalities in lunar motion could be explained by Newton's theory of gravitation. Euler, who as an adjudicator was exempted from participation, drew inspiration from Clairaut's contribution for his own theory, and Mayer eventually became acquainted with both.

Apparently, Mayer's lunar theory did not raise a tremendous interest, judging by the paucity of literature in which *Theoria Lunae* is truly examined. The French astronomer-historian Delambre, although full of admiration for Mayer, is rather short on the contents of *Theoria Lunae*, asserting that

> We will not delve any deeper into his analysis; he warns us himself that one can not see the exactness therein other than by committing oneself to calculations much longer than those that he has made himself,[6]

after which Delambre gives an eight-line summary of Mayer's theory. Forbes, who masterly framed the coherence of Mayer's oeuvre, never ventured too far into the intricacies of lunar motion, and scratched only the surface when it comes to the more technical aspects of Mayer's lunar theory. The only in-depth review of *Theoria Lunae* that I have been able to find is Gautier's,[7] who is of the opinion, writing in 1817, that Mayer's is one of the most elegant and exact lunar theories that have appeared. He, too, asserted that Mayer took Euler's lunar theory of 1753 as a base. Gautier referred to Clairaut as his source of information, and indeed the latter conjectured in 1765 that

> One cannot know if it is by a principle similar to mine, that that skilful Astronomer was the first to reach the simple procedure that he gives for the calculation of his elements, because he has said nothing about his Lunar theory from which he has started out, nor the manner in which he has employed it. I suspect that it is of Monsieur Euler's that he has adjusted the equations through observations, & of which he has made particular good use by thinking of correcting the mean place by the smallest equations before making use of the large ones.[8]

---

[4] Every solution of the differential equations of the sun–earth–moon three-body problem is, of course, only approximative.

[5] Clairaut (1752b).

[6] «*Nous n'entreprendrons pas davantage l'extrait de son analyse; il nous avertit lui-même qu'on ne peut en voir l'exactitude qu'en se livrant à des calculs plus longs que ceux qu'il a faits lui-même*» (Delambre 1827, pp. 443–444). Jean Plana, clearly demonstrating in Plana (1856) that he knew *Theoria Lunae* very well, warned that Delambre was overestimating Mayer.

[7] Gautier (1817, pp. 65–73).

[8] «*On ne peut pas savoir si c'est par un principe pareil au mien, que cet habile Astronome est parvenu le premier au procédé simple qu'il donne pour le calcul de ces élémens, parce*

At that time Mayer's *Theoria Lunae* was still lying in cache, and unfortunately Clairaut did not live to see it. Had he had a chance to see it, he would almost surely have recognized certain aspects of his own lunar theory. Mayer himself, on the other hand, was keen to stress that his theory had nothing of either Clairaut's or d'Alembert's, as appears for instance from a letter to his French colleague Lacaille:

In fact, having worked on the lunar theory since 1749, long before I could have seen the works of those two clever geometers, I managed to obtain rather precise lunar tables, as evidenced by a letter that I have written to Mr. De L'isle around 1749 or 1750. Also the way that I have chosen to resolve the general equations of Mr. Euler is very different from those that Mr. Clairaut and Mr. d'Alembert have decided to follow. An example of this difference can be seen in my printed tables concerning the calculation of the latitude although I have disregarded there some small corrections which I then judged useless. All geometers have only explored the motion of the Nodes and the inclination of the orbit of the Moon separately. Instead, I have extracted the true latitude of the Moon directly from the theory, without requiring either the true place of the Node or the inclination. My complete calculation which I have revised several times is explained in a manuscript that I have sent with the tables to London 3 years ago. I still possess lunar tables that were constructed before the publication of my printed tables, in which all the 22 arguments are specified by the mean motions as in those of Mr. Clairaut; but my equations are always additive. The many equations in these tables have led me to change the form, without losing anything regarding the exactness, in which I have succeeded perfectly. When time will allow me to publish all my researches on the theory of the Moon, it will evidently appear that I have taken nothing from others.[9]

---

qu'il n'a point dit quelle étoit la théorie de la Lune d'où il étoit parti, ni la maniere dont il l'avoit employée. J'ai lieu de croire que c'est de celle de M. Euler dont il a rectifié les équations par les observations, & dont il a tiré un parti singulier en pensant à ne faire usage des grandes équations qu'après avoir corrigé le lieu moyen par les plus petites.» The quote is from the second, enlarged, edition (Clairaut 1752b, 2nd ed., p. 102).

[9] *«En effet, ayant travaillé sur la theorie de la Lune des l'an 1749; longtems avant que j'aye pu voir les ouvrages de ces deux habiles geometres, je fus delors parvenu à des tables de Lune assez exactes, temoin une lettre que j'ai ecrite a M. De L'isle vers l'an 1749 ou 1750. Aussi la route que j'ai tenue pour resoudre les Equations generales de Mr. Euler est elle très differente de celles que M^s. Clairaut & D'alembert ont jugé à propos de suivre. On peut voir un example de cette difference dans mes tables imprimées au sujet du calcul de la latitude quoique j'y aye negligé quelques petites corrections, que je jugai alors inutiles. Tous les geometres ont uniquement cherché separement le mouv. des Noeuds et l'inclinaison de l'orbite de La Lune. Au lieu que j'ai directement tiré de la theorie la latitude vraye de la Lune, sans avoir besoin ni du lieu vrai du Noeud ni de l'inclinaison. Mon calcul entier au quel j'ai retouché plusieur fois est expliqué dans un ecrit que j'ai envoyé avec des tables à Londres il y a 3 ans. Je possede encore des tables de Lune construites avant la publication de mes tables imprimees, dans les quelles tous les argumens au nombre de 22. sont determiné par les mouvemens moyens come dans celles de M^r. Clairaut; mais mes aequations sont toujours additives. Le grand nombre des aequations dans ces tables m'a determiné à en changer la forme, sans perdre quelque chose du cote de l'exactitude; ce qui m'a reussi parfaitement. Lorsque le tems me permettra de publier toutes mes recherches sur la theorie de La Lune, il paroitra evidemment que je n'ai rien emprunté des autres.»* Mayer to Lacaille, post scriptum, 31 Oct. 1758 (Forbes and Gapaillard 1996, pp. 519–520). Forbes identified the letter to De l'Isle that Mayer referred to as the one written on 14 Jan. 1751, from which we quoted earlier (see footnote 9 on Sect. 3.2).

Mayer wrote this in 1758, at a time when the Board of Longitude had still to decide on the awards, a circumstance that may have slightly coloured his remarks. Several points that he raised in the letter are recognizable: for instance, he stressed his characteristic treatment of the latitude equation on several occasions, including *Theoria Lunae* which he had indeed written three years before and which we are about to study. The older lunar tables with their always additive equations must be the kreek version. The form change that Mayer mentioned, refers to his introduction of the multistep format, which will be further investigated in the next two chapters.

5.2 Circumstances of Composition

In the preface of *Theoria Lunae*, Mayer explained that his goal was not to show that the motion of the moon can be accurately derived from Newton's law of gravitation, but rather that nothing *against* his tables could be launched from that side. He mentioned that his theoretical equations in *Theoria Lunae* differ hardly more than half a minute from the tabulated coefficients of his most recent lunar tables at the time (alias wijd), which had been fitted to observed positions of the moon. Mayer explained that his lunar tables are more to be trusted than his lunar theory, since the latter leaves some equations rather inaccurately determined. Moreover, he warned that his lunar theory rests for a part on the solar theory, which still had some uncertainties.[10]

One may wonder: why these excuses, why did he write up the theory anyway? Part of the reason has already been mentioned earlier: Mayer honoured the request of James Bradley, the Astronomer Royal at the time when he applied for the Longitude Prize.[11] Since Bradley would be the prime adviser to the Government, his request could not be dismissed. But though this provides the reason why Mayer wrote *Theoria Lunae*, it leaves the prudent tone unexplained.

Bradley's request for a theory in support of the submitted tables reached Mayer through the usual agents William Best and Johann David Michaelis in November 1754. At that time Mayer was industriously improving the accuracy of his tables by comparison to observations. The tables he was working with were modelled after kil, the tables printed in the spring of 1753. As we will see in Chap. 6, Mayer had no firm theoretical foundation for those tables. We will now take a closer look at the sequence of events of that crucial period.

---

[10] Mayer (1767, pp. 50–51).

[11] Forbes (1980, p. 168). The request was quoted above p. 36.

In the summer of 1753, Mayer had received Euler's *Theoria Motus Lunae*.[12] Mayer had already thanked Euler for his excellent contributions to celestial mechanics, in particular the essay on the great inequality of Jupiter and Saturn, which had shown Mayer how to proceed. He gathered new inspiration from Euler's lunar theory, and in the winter of 1753–1754 he tackled the problem of the lunar latitude, which his tables had up till then represented unsatisfactorily.[13] But his tables for lunar longitude were still at most partially backed up by a theory, therefore Bradley's request for their foundations, later that year, may have embarrassed Mayer. A confession that his tables had developed out of Newton's lunar theory of more than half a century earlier (for indeed this was the case, as I will show in the Chap. 6) would hardly buttress his claim to the Longitude Prize. A sound theory had to be produced, and Mayer set out to his chore. He composed substantial portions of *Theoria Lunae* in January and February of 1755, as is revealed by the title '*Theoria ☽ Jan & Feb 1755*' that he penned down on the cover of a draft of it.[14] Mayer was then still optimistic about his progress. He wrote to Euler on 23 February that the printer Nourse in London would publish the theory when ready.[15] He added that his method was simple, correct, and in excellent agreement with what he had reported in his earlier letter on the problem of the lunar latitude.

But apparently prospects changed soon after. Around that time Mayer's friend and former colleague from Nuremberg, Georg Moritz Lowitz, moved to Göttingen to take up a professorship in Cosmology. We can be sure that the two men discussed the progress of the project. The pressure from London was kept high but Mayer sought excuses to gain more time. The exact reasons for his delay are not known to us. In April, he wrote to Michaelis:

> What pertains to the theory of the tables, and the way in which I have derived them from the law of attraction: I will certainly take trouble in order that I will send it to you at a fitting time when it is brought in convenient order. But I reckon it must be stressed again and again, that this theory, if you perceive something properly, has nothing of any weight to confirm the excellence of the tables and the agreement with heaven itself (although it must be admitted that it has been impossible without the help of the theory to bring the lunar tables to such a degree of perfection as they now rejoice); this will be abundantly clear from that treatise in which I shall shortly expound this theory.[16]

---

[12] Euler (1753).

[13] Forbes (1971a, pp. 69–72, 78, 81–83).

[14] The draft design of *Theoria Lunae* is in Cod. $\mu_{28}^{\sharp}$, which holds three folders of folio papers related to lunar theory. The third of the folders is the item indicated in the text. The second folder is related to *Theoria Lunae* too, the disordered first folder consists of various older attempts at lunar theories.

[15] Forbes (1971a, p. 96).

[16] '*Quod ad theoriam tabularum attinet, et modum quo eas ex lege attractionis derivavi, dabo quidem operam, ut eum in convenientem ordinem redactum suo tempore mittam. Verum iterum iterumque monendum duco, eam theoriam, si quid recte sentis, nihil ponderis habere, ad confirmandum tabularum praestantiam et consensum cum ipso coelo;*

Here, Mayer clearly expressed that the theory had been of help to perfect his tables, but no more than that. How far he had proceeded to bring it in convenient order can be inferred from the fact that in May he was checking the lengthy computations for the theoretical coefficients of the equAtions, treated in Sect. 5.4.9. It is plausible that the difficulty of connecting the theory with the multistepped format of the tables still loomed over the project at that time and that it was the cause of further delay.[17]

Somehow finally Mayer found a way to fulfil the expectations, for he finished *Theoria Lunae*, and in November 1755 the manuscript arrived in the hands of Best and Bradley. Best confessed to Michaelis, through whom the manuscript had been dispatched, that

> The parcel of Prof. Mayer that you sent to me in the second letter, and which concerns the Longitude at Sea, has arrived not exactly at the right moment, because Mylord Anson is making a journey to Bath, and one is presently very engaged at the Admiralty.[18]

Not that Anson would have been able to review *Theoria Lunae*, but certainly his verdict on Mayer's claim was needed, since the Admiralty was considered as having the greatest interest in this matter. Bradley dutifully wrote a positive report soon after he received the manuscript. However, his report addressed the utility of Mayer's lunar *tables* for finding the longitude at sea; regarding the *theory* he wrote only that he had received it.[19]

The Admiralty was indeed busy embarking upon the Seven Years' War (1756–1763), therefore a decision about Mayer's (and Harrison's) contribution to the longitude problem was not reached before 1765, as has been more fully recounted in Sect. 3.7. The theory was finally prepared for the press by Nevil Maskelyne and appeared in 1767, twelve years after its conception.

By that time new methods were tried in celestial mechanics, particularly by Lagrange, and Mayer's theory was already nearly obsolete. Lagrange, primarily concerned with planetary theory, advanced the study of the development of orbital elements over time. In this so-called variation of constants

---

quamvis negari non possit, impossibile fuisse absque eius auxilio tabulas lunares ad hunc perfectionis gradum, quo nunc gaudent evehere; id quod liquebit abunde ex ipsa tractatione, qua hanc theoriam sum expositurus.' Mayer to Michaelis, 14th April 1755, published in Mayer (1770, pp. 43–44). Undoubtably the letter was part of the correspondence over the Longitude Prize.

[17] The changing temper after February is also apparent from the contemporary correspondence between Michaelis in Göttingen and his cousin William Best in London, Michaelis (1794–1796). Of the correspondence between Mayer and Bradley's assistant Gael Morris, only two letters written by Morris in May 1755 and March 1756 could be found (Göttingen, Philos. 159, fol. 26–28).

[18] *"Das in Dero 2tem Schreiben mir zugesandte Päckchen von dem Hrn. Prof. Mayer, die Longitudinem Maris betreffend, ist zwar zu nicht recht gelegener Zeit angelaufen, da Mylord Anson nach Bath verreiset, und man gegenwärtig bey der Admiralität sehr geschäftig ist"*, Best to Michaelis, 13 Sept. 1754, see Cod. Michael. 320, p. 555.

[19] The report is contained in a letter of Bradley to Clevland, Secretary of the Admiralty, dated 10 Feb. 1756; printed in Mayer (1770, pp. cix–cx) and Rigaud (1832, pp. 84–85).

approach, the planet is assumed to move in an elliptic orbit, while the form
and orientation of the ellipse slowly change due to the perturbing forces. Were
these perturbing forces to stop suddenly, then the planet would continue in
an elliptical orbit, defined by the instantaneous values of the orbital elements.
Euler had pioneered the study of the variation of orbital elements, inspired
by a particularly effective device in the lunar theory of Jeremiah Horrocks
(which we will study extensively Chap. 6).

The prime role left for *Theoria Lunae* upon publication, when it was
already twelve years old, would be to show to the world the foundations
of Mayer's lunar tables; a role which it performed particularly poorly, as
Lagrange pointed out to d'Alembert. He thought that Mayer's tables would
lose their good reputation,

> ...if the astronomers, for whom they are destined, were able to judge the theory
> that serves as their foundation. It is remarkable that their author, after having found
> a certain number of equations, rejects some and changes the value of others without
> reason, and one must notice that he makes perpetual changes, because the equations
> of the Tables are not at all the same as the equations corrected by the theory.[20]

The prudent tone of Mayer's preface, alluded to at the beginning of this
section, is a defense against precisely these and similar castigations. To Mayer,
the ostensible accuracy of the tables mattered more than an a posteriori
theoretical support of them.

5.3 Through a Bird's Eye

The problem that Mayer addresses in his theory is now known as the 'main
problem' of lunar motion: he considers the sun–earth–moon system of three
bodies, taken to be point masses. Effects due to the oblateness of the bodies
are not included, although Mayer is aware of their possible impact on the
lunar orbit.[21] He assumes that the motion of the sun relative to the earth is

---

[20] « ...*si les astronomes, pour qui elles sont destinées, étaient bien en état de juger de la
théorie qui leur sert de fondement. Ce qu'il y a de singulier, c'est que l'auteur, après avoir
trouvé un certain nombre d'équations, en rejette les unes et change la valeur des autres
sans raison, et remarquez qu'il y a fait des changements continuels, car les équations des
Tables sont pas tout à fait les mêmes que les équations corrigées de la théorie*» Lagrange to
d'Alembert, 4 April 1771 (Lagrange, 1882, p. 196). D'Alembert, characteristically critical
of just about anyone's work, phrased his verdict as «*Il me semble que comme théorie c'est
assez peu de chose*» ('According to me, this theory is not much') without even having seen
it yet (p. 193 *ibid.*). Not long afterwards, Mayer's theory must have fallen into his hands,
because d'Alembert commented on it in 1773, restricting himself to the differences between
the theory and the fitted tables (d'Alembert 1761–1780, VI, pp. 43–44).

[21] See letters Mayer to Euler, 15 Nov. 1751 and 6 Jan. 1752 (Forbes 1971a, pp. 43, 47),
and my remarks on Forbes' misinterpretation of these letters (p. 32).

known a priori with sufficient accuracy.[22] He takes account of the inclination
of the lunar orbit (i.e., the angle between the lunar orbit and the ecliptic)
and makes use of the small size of this angle. Not included in Mayer's lunar theory are the small perturbations caused by other planets, and secular
acceleration. In other words, the theory seems to be a state-of-the-art mid-
eighteenth century one.

Now follows a brief general outline of Mayer's lunar theory, intended to
provide a skeleton to the more extensive study in the next section, up to
the point where Mayer introduced the multistepped format. The numbers
between square brackets correspond in both sections, so that the reader may
always return here to regain a lost thread.

[1] Mayer sets up a spherical coordinate system (Fig. 5.1) with the origin in
the centre of the earth, and the plane of the ecliptic as the plane of reference.

[2] He finds expressions for the gravitational accelerations that the moon,
earth, and sun exert on each other. With Newton's second law on the accelerations caused by forces, he arrives at three differential equations expressing
the lunar acceleration in the radial, tangential, and axial directions.

[3] These equations have time as independent variable, for which Mayer
substitutes the lunar mean motion. This eliminates the mean earth–moon
distance.

[4] He expresses the sun–moon distance in the coordinates of those bodies.

[5] Mayer integrates the equation for the tangential component of the
acceleration once. In the simpler two-body problem, this equation has a zero
right-hand side, whence it induces the area law. But in the three-body case an
unevaluated integral of the gravitational accelerations remains on the right-
hand side after the integration.

Up to this point Mayer's theory is very similar to Euler's lunar theory of
1753, and partly also to the latter's essay on the great inequality in the movement of Jupiter and Saturn. Then Mayer makes three more substitutions.

[6] Two of the substitutions can already be found in Euler's theory so they
were not new.

[7] The third substitution signifies a break with Euler's line of development.
It introduces a new independent variable, which is somewhat related to the
true anomaly in an osculating orbit. Besides, Mayer manipulates the equation
with the latitude coordinate, which Euler had preferred to split into one for

---

[22] This is a reasonable assumption; from observations it was known that the apparent orbit
of the sun around the earth is not sensibly perturbed by the moon. The largest perturbation
in the Keplerian apparent orbit of the sun (larger than any of the planetary perturbations),
is caused by the circumstance that the barycentre of the earth–moon system, rather than
the centre of the earth, takes its place in the focus of the ellipse. But at any time the
barycentre is located well below the earth's surface. Therefore, the perturbation in the
apparent motion of the sun is smaller than the angle subtended by the earth's radius as
seen from the sun, i.e., smaller than the solar parallax. The magnitude of the latter quantity
was not well known. By the beginning of the eighteenth century its value was believed to
be about $10''-12''$ (now $8.8''$), well below the threshold of observational accuracy.

inclination and one for node position, into a form nearly identical to the equation for the moon's longitude, so that he can solve it later with the same technique.

[8] After these manipulations, latitude occurs still in one place outside the latitude equation. But because the inclination of the lunar orbit is only about 5°, Mayer contends to take an estimation of latitude in that place, so that he makes solving of the latitude equation an independent problem.

[9] At this stage, Mayer has formulated the lunar motion problem as five ordinary differential equations. Three of these equations specify the response of the moon to gravitational accelerations. Two of these three equations are of second order and of similar structure; one equation is of first order. The remaining two equations are also of first order: one of them relates the independent variable to the lunar *true* longitude, the other relates it to the lunar *mean* longitude.

[10] Remarking that it is apparently hopeless to try to solve these equations exactly, Mayer then introduces trigonometric series with indeterminate coefficients to approximate a solution. He does not specify that the periods of the arguments of the series have rational ratios to each other, and in fact the frequencies of the terms could be incommensurable. Therefore, he neither intended nor implied a periodic solution.

[11] A good deal of rather dull work is then necessary to express every term that occurs in the differential equations as trigonometric series too.

[12] His next task is to determine the coefficients of the series. At this stage Mayer draws quite heavily on his own earlier results of mixed (theoretical as well as observational) provenance.

[13] When these coefficients have been found, he has the motion of the moon expressed as a function of an independent variable that is inappropriate for practical work. Therefore, Mayer reformulates the solution in terms of the lunar mean longitude by means of a series inversion.

At that point, we break off our discussion of *Theoria Lunae*: the solution of the equations of motion is obtained in the single-stepped format. But Mayer's tables employ a multistepped scheme, and how Mayer accommodates the difference between the two in *Theoria Lunae* is the subject of Chap. 7. Mayer also imparted that his tables had been adjusted to observations, but his methods to do so were not disclosed in *Theoria Lunae*; they are the subject of Chap. 8. Mayer's evaluation of the *solar* parallax based on the adjusted coefficient of the so-called parallactic equation of the moon, though interesting in itself, is outside the scope of the current research.[23]

---

[23] See, e.g., Mayer (1767, p. 53), Laplace (1802, p. 657), Godfray (1852, pp. 79–81).

5.4 Mastering the Theory

5.4.1 Notation

In order to write understandable mathematical formulae, one has to adhere to established conventions, but the conventions may change over time. A historian of mathematics, being an intermediary between past and present times, tries to strike a balance (depending on the specific needs of his message) between form and content of the mathematics that he passes on. My current aim is to convey the contents of Mayer's lunar theory; therefore, I have deviated from Mayer's choice of symbols where confusion was likely to arise. Particularly the letters π and e now have a definite mathematical connotation which was virtually absent in the mid-eighteenth century. I have also modernized Mayer's notation in other respects. For example, the formula $dp = ex^2\, dq$ in *Theoria Lunae* is rendered here as $dq/dp = 1/hx^2$, avoiding loose differentials and the symbol e. Displays 5.1 and 5.2 list the most important constants and variables of the theory. The numbers in square brackets in the following text correspond with the same numbers in Sect. 5.3.

[1] We start our discussion of *Theoria Lunae* with the geometry of the sun–earth–moon system as depicted in Fig. 5.1. Let T be the earth, S the sun, and L the moon, all three considered as point masses. L is projected perpendicularly onto A in the plane of the ecliptic ΥST; this is in the plane of the drawing, and dotted lines are outside it. $T\Upsilon$ is a fixed direction in this plane. Mayer remarks that the plane of the ecliptic is not exactly fixed, but that its motion is negligible for his purpose.[24] Likewise, Mayer uses the *symbol* Υ, but he is careful *not* to call it the vernal equinox, which is not in a fixed direction due to precession and nutation.

Notwithstanding these details, he takes the angle $\varphi = \angle \Upsilon T A$ as the longitude of the moon; $l = \angle LTA$ its latitude; and $av = AT$ its curtate distance,[25] where the constant a denotes the mean distance of the moon from the earth and $v \approx 1$ is a variable. We may interpret the triple (v, φ, l) as the spherical coordinates of the moon relative to the earth, albeit in a somewhat modified form because the scaled curtate distance takes the place of the radius coordinate.

To represent the sun in the three-body geometry, let k be the ratio of the mean moon–earth distance a to the mean sun–earth distance. Mayer's value for k agrees to his initial estimate of $10.8''$ for the solar parallax. The mean sun–earth distance is then a/k. Let the true sun–earth distance be given by $ST = ay/k$, which defines the variable y, and let the true sun–moon distance

---

[24] This is even more the case since the lunar orbit tends to remain fixed with respect to the moving ecliptic, as Laplace pointed out (Laplace 1802, p. 374); his proof takes several pages.

[25] The *curtate distance* is the distance projected perpendicularly onto the plane of the ecliptic.

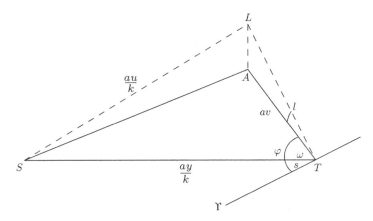

Fig. 5.1 Parameters in the lunar problem

Display 5.1 Constants with the values that Mayer assumed

Constant	Value	Meaning
a		Mean lunar distance, to be eliminated in [4]
A	0.0545400	Mean lunar eccentricity
ϵ	0.0168350	Solar eccentricity
k	0.0031505	Ratio of (mean) solar and lunar parallaxes
n	0.0748017	Solar mean motion
α	0.9915478	Lunar mean motion relative to the apogee
r	0.9251983	Mean motion relative to the sun $(r = 1 - n)$
i	1.0040194	Mean motion relative to the node
g	1.018408	See page 70
h	1.0047450	See page 80
τ	0.9999183	
f^2	0.9958270	

The mean motions are stated as ratios to the lunar mean motion. h, τ and f^2 are composite (see text)

be $SL = au/k$, which defines u. Like v, the variable y is rather close to unity. Moreover, u is close to unity, too, because the moon is always much closer to the earth than to the sun. Further, let the longitude of the sun be denoted by $s = \angle \Upsilon TS$ and the difference of the lunar and solar longitudes by $\omega = \varphi - s$.

5.4.2 Differential Equations in Spherical Coordinates

[2] Euler had been the first to formulate Newton's second law as differential equations, equivalent to

$$\frac{d^2x}{dt^2} = \frac{F_x}{2M}, \quad \frac{d^2y}{dt^2} = \frac{F_y}{2M}, \quad \frac{d^2z}{dt^2} = \frac{F_z}{2M}, \tag{5.1}$$

Display 5.2 Variables

Variable	Meaning
φ	True longitude of moon
l	True latitude of moon
v	Scaled curtate distance of moon
x	$x = 1/v$
y	Scaled true earth–sun distance
u	Scaled true moon–sun distance
s	True longitude of sun
σ	True anomaly of sun
ω	$\omega = \varphi - s$
q	Mean longitude of moon
ψ	Mean longitude of sun
X,Y,Z	Forces in the right-hand sides of eqns. (I), (II), (III) (see p. 71)
P	$P = \int Y x^{-1}\, \mathrm{d}q$
p	Defined by $\mathrm{d}p/\mathrm{d}q = hx^2$, $p(0) = q(0)$

for each of the three rectangular coordinates x, y, z of a body, with forces F_x, F_y, F_z and mass of body M. Quite some work is needed to transform these equations from rectangular to spherical coordinates. Euler had shown how to do it,[26] and Mayer felt apparently no need to repeat Euler's calculations in *Theoria Lunae*. Mayer formulated (5.1) in rectangular and spherical coordinates and referred to Euler for the details.

Thus Mayer arrives at the differential equations of motion of the moon in spherical coordinates, subject to yet unspecified accelerations F_r parallel to AT, F_t perpendicular to AT and parallel to the ecliptic plane, and F_a perpendicular to the ecliptic. In the radial, tangential, and axial directions, these differential equations are, respectively:

$$\frac{\mathrm{d}^2 v}{\mathrm{d}t^2} - v\left(\frac{\mathrm{d}\varphi}{\mathrm{d}t}\right)^2 = -\frac{1}{2a}F_r,$$

$$2\frac{\mathrm{d}v}{\mathrm{d}t}\frac{\mathrm{d}\varphi}{\mathrm{d}t} + v\frac{\mathrm{d}^2\varphi}{\mathrm{d}t^2} = -\frac{1}{2a}F_t, \qquad (5.2)$$

$$\frac{\mathrm{d}^2(v\tan l)}{\mathrm{d}t^2} = -\frac{1}{2a}F_a.$$

No approximations have been made in arriving at these equations. Mayer follows Euler's practice in the use of the factor $1/2$, which makes for a simpler relation $v^2 = h$ between speed v and distance traversed h for a freely falling body.[27]

---

[26] Euler (1749a, p. 54), Euler (1749b, pp. 9–12), Euler (1753, §3–6).

[27] See Euler (1749b, §19) (and the foreword by Max Schürer of the reprint), also Volk (1983, p. 346), Wilson (1985, p. 72). A note in Nevil Maskelyne's papers, RGO 4/187[21], reveals that the latter was for a time somewhat confused by this practice.

Next, Mayer specifies the accelerations, although he calls them forces. Let the masses of sun, earth, and moon be $M_\odot$, $M_\oplus$, M_D. Using Newton's law of gravitation, Mayer finds that:

along LT, the acceleration equals $\dfrac{(M_\mathrm{D} + M_\oplus)\cos^2 l}{a^2 v^2}$,

along LS, it is $\dfrac{M_\odot k^2}{a^2 u^2}$, and

along ST, $\dfrac{M_\odot k^2}{a^2 y^2}$.

We observe that, according to Newton's law of gravitation, the lunar and terrestrial masses should be added to the solar mass in the second and third expressions, respectively. The simplification made by omitting them leads to a negligible error.

By decomposition, the forces in the right-hand sides of (5.2) become:

$$F_\mathrm{r} = \frac{(M_\mathrm{D} + M_\oplus)\cos^3 l}{a^2 v^2} + \frac{M_\odot k^3 v}{a^2 u^3} + \left(\frac{M_\odot k^2}{a^2 y^2} - \frac{M_\odot k^2 y}{a^2 u^3}\right)\cos\omega,$$

$$F_\mathrm{t} = -\left(\frac{M_\odot k^2}{a^2 y^2} - \frac{M_\odot k^2 y}{a^2 u^3}\right)\sin\omega, \qquad (5.3)$$

$$F_\mathrm{a} = \frac{(M_\mathrm{D} + M_\oplus)\cos^3 l \tan l}{a^2 v^2} + \frac{M_\odot k^3 v \tan l}{a^2 u^3}.$$

Analogous expressions can be found in Euler's treatises.[28]

5.4.3 Two Eliminations

Next Mayer rescales his time variable and he eliminates the distance u from (5.2). The benefit of the time rescaling is that it eliminates the earth–moon distance a, which was not yet very accurately known then. The rescaling is effectuated as follows.

[3] In the autonomous differential equations (5.2), time appears only in the form of the differential dt and its square. Probably inspired by Euler,[29] Mayer replaces the time by the mean ecliptic longitude of the moon. Let therefore q be the mean longitude of the moon at any moment t, and let ψ be the mean longitude of the sun at that time. Their derivatives dq/dt and $d\psi/dt$ are the mean motions of moon and sun, which are constants very accurately known from observations. Kepler's third law in analytical form for the motion of the earth around the sun reads

---

[28] E.g., Euler (1749a, p. 56) and Euler (1753, §§15–20).
[29] See Euler (1749b, §26).

$$\left(\frac{\mathrm{d}\psi}{\mathrm{d}t}\right)^2 = \frac{M_\odot + M_\oplus}{2(a/k)^3}. \tag{5.4}$$

Putting n for the ratio $\mathrm{d}\psi/\mathrm{d}t : \mathrm{d}q/\mathrm{d}t$ of the solar and lunar mean motions,[30] Mayer obtained

$$\left(\frac{\mathrm{d}q}{\mathrm{d}t}\right)^2 = \frac{k^3(M_\odot + M_\oplus)}{2n^2a^3}, \tag{5.5}$$

which suffices to replace the differential of the time variable t by that of the new time-like variable q. When subsequently (5.2) and (5.3) are rewritten, the factor a^3 will be seen to cancel in the right-hand sides. The factor 2 in the numerator arises from Euler's practice referred to above on p. 68.

For later reference, we note that Mayer puts the constant[31]

$$g = \frac{n^2(M_\mathrm{D} + M_\oplus)}{k^3(M_\odot + M_\oplus)}.$$

We also remark that differential equation (5.5) relates q to t, but that an initial condition has not been given. Therefore (5.5) defines q only partially. A solution of the differential equation would be of the form $q(t) = c_1 t + c_2$ (modulo 360°), and implicitly Mayer chooses the boundary condition in such a way that $c_2 = 0$.

[4] Now we turn to the elimination of u, the scaled distance between sun and moon. In triangle ATS we have $AT = av$, $TS = ay/k$, and $\angle ATS = w$,[32] so Mayer can express SA by

$$SA^2 = a^2v^2 + \frac{a^2y^2}{k^2} - \frac{2a^2vy}{k}\cos w.$$

In the right-angled triangle SAL we have $SL^2 = SA^2 + a^2v^2\tan^2 l$; therefore, applying $1 + \tan^2 l = \cos^{-2} l$,

$$\left(\frac{au}{k}\right)^2 = SL^2 = \frac{a^2v^2}{\cos^2 l} + \frac{a^2y^2}{k^2} - \frac{2a^2vy}{k}\cos w,$$

so that

$$u = \sqrt{y^2 - 2yvk\cos w + \frac{v^2k^2}{\cos^2 l}},$$

where a has fallen out of the equation again. Using the binomial theorem, he obtains the expansion

---

[30] Actually Mayer defines this as $\bar{n}$, and he defines $n^2 = M_\odot\bar{n}^2/(M_\odot + M_\oplus)$; but the mass $M_\oplus$ is very small compared to the mass $M_\odot$, so $\bar{n} \approx n$, and Mayer mixes the two at will. I will not distinguish between $\bar{n}$ and n.

[31] In §8 of both the manuscript and the printed edition of *Theoria Lunae*, the square on the n is missing.

[32] Mayer's text (§9) has SAT instead of $\angle ATS$.

$$\frac{1}{u^3} = \frac{1}{y^3} + \frac{3kv\cos\omega}{y^4} + \frac{15k^2v^2(1+\cos 2\omega)}{4y^5} - \frac{3k^2v^2}{2y^5\cos^2 l}, \quad (5.6)$$

up to the second order in k. Mayer then substitutes (5.6) in (5.3). In the end result, Mayer keeps terms up to the first order in k. Apparently he has put $\cos l = 1$ here, equivalent to neglecting the difference of SA and SL right from the beginning (the difference is less than $1 : 4 \times 10^7$).

After these substitutions and rewriting, Mayer arrives at the following equations to describe the motion of the moon:[33]

$$\frac{d^2 v}{dq^2} - v\left(\frac{d\varphi}{dq}\right)^2 = -\frac{g\cos^3 l}{v^2} + \frac{n^2 v}{2y^3} + \frac{3n^2 v\cos 2\omega}{2y^3}$$
$$+ \frac{9kn^2 v^2\cos\omega}{8y^4} + \frac{15kn^2 v^2\cos 3\omega}{8y^4}, \quad (I)$$

$$2\frac{dv}{dq}\frac{d\varphi}{dq} + v\frac{d^2\varphi}{dq^2} = -\frac{3n^2 v\sin 2\omega}{2y^3} - \frac{3kn^2 v^2\sin\omega}{8y^4} - \frac{15kn^2 v^2\sin 3\omega}{8y^4}, \quad (II)$$

$$\frac{d^2(v\tan l)}{dq^2} = -\left(\frac{g\cos l^3}{v^2} + \frac{n^2 v}{y^3} + \frac{3kn^2 v^2\cos\omega}{y^4}\right)\tan l. \quad (III)$$

For brevity, the right-hand sides of (I), (II), and (III) are denoted by $-X$, $-Y$, and $-Z\tan l$, respectively. These are the gravitational accelerations up to and including the first order in k.

It is interesting to note in what respect these equations differ from the easier two-body problem. When two bodies revolve around each other, the motion is contained in a plane, and coordinates can be chosen to reflect this, obviating the need for the third differential equation for latitude. Apart from that, we may note that the forces due to the perturbing influence of the sun are present in the terms involving y. Were the sun's influence to stop, then only the terrestrial influence present through the two terms involving gv^{-2} in (I) and (III) would remain in the right-hand sides. In particular, the right-hand side of the second equation is zero in the case of two bodies. This property induces the law of equal areas, also known as Kepler's second law, which is apparently violated in the three-body case.

The terms through which the solar perturbation is expressed all have n^2 or kn^2 as a factor. Considering that $n^2 \approx 0.005$ and $k \approx 0.003$, we note that k and n^2 are of the same order of magnitude. It is useful in lunar theory to distinguish between quantities $\approx 1/20$ of the *first order*, those $\approx (1/20)^2$ of the *second order*, etc.[34] Following this terminology, n is a quantity of the first order, while k and the perturbing solar accelerations $(\sim n^2)$ are quantities of

---

[33] These equations occur twice in §10 of *Theoria Lunae*, with some typographical errors which do not propagate into the sequel.

[34] Laplace (1802, pp. 356, 387), Godfray (1852, pp. 19, 23). According to Gautier (1817, p. 40), this distinction is due to d'Alembert.

the second order. We see therefore that the terms involving kn^2 included in Mayer's equations are of the fourth order, comparable to the square of the perturbing acceleration.

5.4.4 An Inventive Substitution

Mayer says that a solution of these equations is best found by means of approximation (it is now known to be the only way). He proceeds as follows.

[5] He multiplies the second equation by v, then he integrates once, and obtains:

$$\frac{v^2 \, d\varphi}{dq} = h - \int Yv \, dq, \qquad (5.7)$$

where h denotes a constant to be determined later, absorbing the constant part of the unevaluated integral; h is twice the mean rate at which the earth–moon radius sweeps out area. Clairaut had first applied this integration in about 1745.[35]

[6] Mayer then puts

$$x = \frac{1}{v} \quad \text{and} \quad P = \int \frac{Y}{x} dq. \qquad (5.8)$$

The first of these is a rather natural substitution to make, because in the two-body problem it renders the differential equations in a linear form. The second is little more than a shorthand notation for an unevaluated integral, similar to practices of Euler and Clairaut.

[7] The next substitution, introducing Mayer's new independent variable p, is

$$\frac{dq}{dp} = \frac{1}{hx^2}. \qquad (5.9)$$

This substitution is a characteristic feature of Mayer's lunar theory.[36] We note that p is not uniquely defined because an initial value is not given, just as was the case with q before; it is implied that $p = 0$ when $q = 0$.

With these three substitutions, (5.7) turns into

$$\frac{d\varphi}{dp} = 1 - \frac{P}{h}, \qquad (5.10)$$

---

[35] Wilson (1985, p. 79).

[36] It might seem more natural to represent the substitution as $dp/dq = hx^2$, since p is the new variable; however, this would obfuscate the further description of Mayer's theory, in particular in Sect. 5.4.8 below. Mayer wrote the three substitutions as $v = 1/x$, $\int Y/x \, dq = P$, and $hx^2 \, dq = dp$, adding an alternative form $dq = dp/hx^2$ for the latter (Mayer 1767, §14).

while differentiation of P yields

$$\frac{\mathrm{d}P}{\mathrm{d}p} = \frac{Y}{hx^3}. \tag{II$'$}$$

These two equations together replace (II); subsequently Mayer rewrites (I) and (III) also in the new variables. The first equation becomes

$$\frac{\mathrm{d}^2 x}{\mathrm{d}p^2} + x - \frac{X}{h^2 x^2} - \frac{2Px}{h} + \frac{P^2 x}{h^2} = 0. \tag{I$'$}$$

He transforms the remaining third equation as follows. After applying the substitutions, he adds the transformed first equation (I$'$) multiplied by $\tan l$, divides the sum by x, and arrives at

$$\frac{\mathrm{d}^2 \tan l}{\mathrm{d}p^2} + \tan l + \frac{(Z - X)\tan l}{h^2 x^3} - \frac{2P\tan l}{h} + \frac{P^2 \tan l}{h^2} = 0. \tag{III$'$}$$

[8] The latitude l occurs not at all in (II$'$), and it occurs only once in (I$'$) where it is present in a term $-gh^{-2}\cos^3 l$ included in X. But because $|l| < 5\frac{1}{2}°$, this term is not very sensitive to errors in l. Therefore, Mayer feels justified to use an independently obtained estimate of l in (I$'$). Consequently, he treats the system of three differential equations as two independent subsystems: one comprising equations (I$'$) and (II$'$), which will be solved first, and the other with only the transformed latitude equation (III$'$). The structure of (III$'$) is very much alike (I$'$), and he will ultimately solve it by the same techniques.

In contrast, Euler's and Clairaut's policy was to split the latitude equation into two equations of first order: one describes the motion of the node, the other the inclination of the orbit. Mayer stresses that their policy not only leads to more work, but also leads to tables that are less useful to astronomers. Indeed, Euler and Clairaut failed to equal the economy of method that Mayer accomplished. Although the node and inclination of the moon certainly do have their merit for astronomers, especially in the prediction of eclipses, they would more often be just clumsy intermediates in the computation of the lunar latitude. We will not pursue the solution of the latitude equation.

[9] Now we may summarize Mayer's position as follows. He has the motion of the moon described by the three differential equations (I$'$), (II$'$), and (III$'$); the first and last of these are of second order and of similar form, the middle one is of first order. These equations have an independent variable that is linked to both true and mean longitude of the moon by two subsidiary equations (5.9) and (5.10). The original equations (I), (II), and (III) each contained derivatives of more than one coordinate, but the substitutions (primarily (5.9)) have separated them. Although the equations have not been decoupled, they do look less complicated this way.

5.4.5 Interpretation of p

The two subsidiary equations (5.9) and (5.10) link the new independent variable p to the true longitude φ and mean longitude q of the moon. This independent variable is a characteristic feature of Mayer's theory and it is worth to meditate over its meaning for a while. For that purpose we first turn to the two-body case again; the discussion of *Theoria Lunae* will be continued in Sect. 5.4.7.

It is a standard result in the study of the two-body problem that, with θ for true anomaly, $v^2 d\theta/dt = $ constant. This result is already comprised in (5.7). To see this, observe that (1) dq/dt is constant (it is the mean motion of the moon); (2) in absence of perturbing masses, the moon would move in a stationary elliptic orbit, so that its true longitude φ and true anomaly θ differ only by a constant, the longitude of apogee; and (3) $Y = 0$ in absence of perturbing masses. Still with perturbing masses absent, the variable v (or equally $1/x$) would be the scaled earth–moon radius in an elliptic lunar orbit. Thus, the difference of $v^2 d\theta/dt$ and $v^2 d\varphi/dq$ is constant. Now the new variable p is introduced by the differential equation (5.9), which may also be written as $v^2 dp/dq = h$; therefore in the two-body case p would denote true anomaly, except perhaps for a constant phase difference.

Now take the solar perturbation into account again, and observe that v is actually the *perturbed* earth–moon distance. Clearly, then, p can no longer be identified as the true anomaly in an *un*perturbed elliptic orbit.

A somewhat (but not entirely) similar idea is incorporated in a memoir of Euler's whose title began *Nouvelle méthode de déterminer les dérangemens dans le mouvement des corps célestes* and which appeared in print in 1770, but was already presented for the Academy of Berlin in 1763. There, Euler arrived at a relation equivalent to, in our notation,

$$\frac{dq}{d\varphi} = \frac{1}{hx^2},$$

with the understanding that h is variable instead of a constant.[37] In this particular memoir, Euler explored ways to describe small perturbations of elliptical orbits in general. The same relation appeared again in another of his papers where he addressed the lunar theory in particular; there he discussed what is now known as the *osculating orbit*, i.e., the orbit that the moon would follow if the sun's action were suddenly stopped at any moment.[38]

This notion of the osculating orbit is closely associated with the method of the variation of constants, to which Lagrange contributed much. In

---

[37] Euler's expression reads $vv\, d\varphi = r\, dt$; the equivalence follows from $x = 1/v$ and $q = \text{constant} \cdot t$ (Euler 1770a, p. 173).

[38] *«si la force perturbatrice du Soleil venoit à évanouir subitement»* (Euler 1770b, p. 185); two pages further on Euler substitutes $vv\, d\varphi = s\, dt$, where s has the same meaning as r has in Euler (1770a) (footnote 36).

an unperturbed elliptic orbit the six orbital elements are constants; if perturbations are present, it is often feasible to study the motion in an orbit that is thought of as a slowly form-changing ellipse, considering the orbital elements no longer as constants but as functions of time. When for any desired time the instantaneous elements of the orbit are known, then these may serve to compute the position of the body in the usual way for unperturbed motion.

The difference between Mayer's approach and the approach of Euler in the articles just mentioned is primarily that Euler intended to study perturbations through the slow changes in the elements, whereas Mayer continued straightforwardly to the computation of the moon's coordinates. Also, contrary to Mayer, Euler regarded h as a variable quantity which expressed the instantaneous angular momentum in the osculating orbit, so that his φ can be termed the osculating true anomaly. Mayer did not enunciate any speculations about the nature of his independent variable, and the sequel of his theory shows that he did not consider it in connection with variation of constants. The reason for him to introduce p was, presumably, that it rendered the differential equations in a particularly pretty form.[39]

Returning now to a comparison of the lunar theories: Euler's and Mayer's theories run largely in parallel up to the point where Mayer introduces this new independent variable. Euler continued by rewriting his differential equations in a form which takes advantage of the relatively small deviations of lunar motion with respect to an elliptical orbit, with true anomaly in the unperturbed orbit as the independent variable.[40] From here onwards the theories of Euler and Mayer develop rather dissimilarly.

5.4.6 Intermezzo: A Link to Clairaut?

A strong but not immediate obvious relation exists between Mayer's variable p and certain expressions in Clairaut's theory, as I will now demonstrate. In Clairaut's theory,[41] (5.7) takes the form

---

[39] p is certainly not eccentric anomaly. Proof (see any textbook on elementary celestial mechanics for the formulae): assume an elliptical orbit, let E be the eccentric anomaly and ϵ the eccentricity. We have $E - \epsilon \sin E = q$; differentiate with respect to q to get

$$\frac{\mathrm{d}E}{\mathrm{d}q}(1 - \epsilon \cos E) = 1.$$

But $1 - \epsilon \cos E = 1/x$, the ratio of the true to the mean distance of the moon from the earth. Therefore, for the eccentric anomaly we have $\mathrm{d}E/\mathrm{d}q = x$, while p obeys $\mathrm{d}p/\mathrm{d}q = hx^2$, so $p \neq E$. Forbes and Wilson (1995, p. 64) assert that Mayer relates both true and mean anomaly to eccentric anomaly; and that he finally eliminates the latter. Their statement can only be interpreted to imply that p is eccentric anomaly, but I have just proved that this is false.

[40] Euler (1753, Chap. 3).

[41] Clairaut (1752b); substantial portions of it also appeared in Clairaut (1752a).

$$v^2 \frac{\mathrm{d}\varphi}{\mathrm{d}t} = f + \int \Pi v \, \mathrm{d}t. \tag{5.11}$$

Here, f and Π are virtually the same as Mayer's h and Y except for a constant factor relating time and mean longitude. Clairaut multiplies both sides by $\Pi v \, \mathrm{d}t$ and integrates again to get

$$\int \Pi v^3 \, \mathrm{d}\varphi = f \int \Pi v \, \mathrm{d}t + \tfrac{1}{2} \left(\int \Pi v \, \mathrm{d}t \right)^2.$$

Completing the square on the right-hand side and extracting roots, he obtains

$$f + \int \Pi v \, \mathrm{d}t = \sqrt{f^2 + 2 \int \Pi v^3 \, \mathrm{d}\varphi}.$$

He substitutes this result in (5.11) and arrives at

$$v^2 \frac{\mathrm{d}\varphi}{\mathrm{d}t} = f \sqrt{1 + 2\varrho}, \tag{5.12}$$

where

$$\varrho = \frac{1}{f^2} \int \Pi v^3 \, \mathrm{d}\varphi.$$

Clairaut used this result to substitute true longitude for time, where Mayer had substituted mean longitude using (5.5). But, more interestingly, if we express (5.7) partly in Mayer's new variables and then combine with (5.10), we obtain

$$\frac{v^2 \, \mathrm{d}\varphi}{\mathrm{d}q} = h - P = h \frac{\mathrm{d}\varphi}{\mathrm{d}p}.$$

This implies that Mayer's $\mathrm{d}\varphi/\mathrm{d}p$ is proportional to Clairaut's $\sqrt{1 + 2\varrho}$.

It is intriguing that Mayer chose an independent variable with an unfamiliar astronomical interpretation, which is yet so strongly related to certain aspects of Clairaut's theory. There is some evidence, inconclusive however, that Mayer conceived the substitution (5.9) quite late, perhaps as late as the beginning of 1755. He possessed Clairaut's theory already at least half a year then, so it is possible that Mayer's choice of independent variable was inspired by it.

5.4.7 Trigonometric Series

[10] We now continue our investigation of Mayer's *Theoria Lunae*. From here on, the manipulation of the differential equations makes place for the handling of trigonometric series. With p as the independent variable, Mayer needs to express every term in the differential equations (particularly x, y, l, ω, and P)

in this p, developing in series expansions where necessary. To this end, he asserts that x and P can be expressed by trigonometric series:

$$x = 1 - A \cos \alpha p - B \cos \beta p - C \cos \gamma p - \cdots , \qquad (5.13)$$

$$\frac{\mathrm{d}P}{\mathrm{d}p} = a \sin \alpha p + b \sin \beta p + c \sin \gamma p + \cdots . \qquad (5.14)$$

Trigonometric series were appealing because it was known from experience that the motion of the moon is nearly periodic.[42] We pause to ascertain whether the series as Mayer proposed them are indeed appropriate. The constant term 1 in (5.13) is in accord with the distance scaling as explained in Sect. 5.4.1. Further, it can be readily seen that Y in (II), (II$'$) consists of sine terms; Y and Yx^{-3} are thus both odd functions. Therefore the null constant in (5.14) is in harmony with equation (II$'$). Besides, X in equations (I), (I$'$) contains cosine terms. Thus, after plugging in the above series, all terms that appear in (II$'$) are odd functions and all terms in (I$'$) are even. The series that Mayer proposed are therefore appropriate. Moreover, we can see that they extend the two-body case (where $x = 1 - A \cos p$ and $\mathrm{d}P/\mathrm{d}p = 0$) with additional perturbational terms.

Now let us look forward and see how those series will serve him. Equation (5.14) integrated once yields a cosine series for P. Substitution of this series in (5.10), followed by integration, yields a sine series development for φ. This series, (5.13), and the solution of the latitude equation (III$'$) to be derived separately, together express the true position (longitude and latitude) of the moon for any value of p. Unfortunately, p is a rather awkward quantity for practical purposes. Therefore Mayer re-expresses the series using q as the independent variable, after expressing p in terms of q by integration of (5.9) and inversion of series.

The programme will unfold as follows. Section 5.4.8 shows how Mayer expresses the remaining variables y, l, ω in (I$'$) and (II$'$) through p. Section 5.4.9 then explains how he derives the values of the coefficients B, C, $\ldots$ from (I$'$), the values of b, c, $\ldots$ from (II$'$), and those of α, β, γ, $\ldots$ from both. The latter are clearly related to the periods of the perturbations. Special considerations apply to A and α, as will now be explained.

We know already (see Sect. 5.4.3) that q is the lunar mean motion, and that $q = p$ if solar perturbation ceased: in that case (5.13) would reduce to $x = 1 - A \cos \alpha q$, describing the radius vector in an elliptic orbit. Hence it follows that A is equal to the (mean) eccentricity of the lunar orbit, which is not obtainable from the differential equations and must take a value derived from observations. Also, αq is lunar mean anomaly and $(1 - \alpha)q$ is the mean motion of lunar apogee: in absence of perturbations we would have $\alpha = 1$. Mayer considers (undoubtedly encouraged by the recent success of Clairaut in his derivation of the apogee motion) that if his theory yields a value for α

---

[42] Cf. fn. 27 in Sect. 6.4. Besides, note that the letter a, which originally denoted the mean distance of the moon, is reused here with a different meaning.

close to the value known from observations, then this will be a confirmation of the Newtonian law of gravitation. Yet, as we will see, α appears in many more places in the theory. In those other places it is always multiplied by a small constant, and thus much less significant, therefore Mayer allows to save himself a lot of labour by taking an empirical value for α in those cases.[43]

We should note some particulars about the series (5.13) and (5.14). First, Mayer accomplished a surprising efficiency in the number of terms that he explicitly wrote down. The indeterminates β and γ will prove to be necessary and sufficient to explore a myriad of trigonometric terms that arise in the sequel. With just three explicit terms, Mayer keeps his formulae manageable.[44] This is a characteristic feature of his theory. Euler, also applying trigonometric series in his lunar theory,[45] consumed alphabets at high speed: upper and lower case, primed and unprimed, Roman and Fraktur, in want of identifiers for his terms.

In the second place, regarding trigonometric series and periodicity, we remark that Mayer was very well aware that the moon's motion was almost, but not exactly, repeated after a Saros of 223 lunations. The non-periodicity of lunar motion implies that the argument parameters α, β, γ, ... do not all have rational proportions. Mayer did not consider the questions of rationality and periodicity.

Lastly, it needs to be remarked that Mayer integrated several series term by term. In a modern theoretical context, an investigation into the validity of the integrations would be in order. But in the eighteenth century such an act was highly exceptional, and it is also neglected in many modern applications.

---

[43] The full text, which was paraphrased above, reads:. 'Thus it has to be observed that the letter α is as much as unknown and indeterminate, at least in this first term; in the rest however, where it is combined with others, it will be possible to suppose its true value from observations, in order that the determination of the remaining quantities will come out so much more easy and exact. Now when, after the troubles to determine this are resolved, the same value for the letter α comes out of the first term of equation I that the observations show, then this will be evidence that the Newtonian theory of gravity, where the forces of the Sun, the Earth and the Moon are fixed proportional to the squares of the distances reciprocally, is the truth and according to observations; if not, while the calculation is nevertheless accurately performed, it will inform us that it [i.e., Newton's theory] needs a correction (*'Spectari ergo debet haec litera α tanquam incognita aut indeterminata, saltem in hoc primo termino; in reliquis tamen, ubi ea cum aliis [...] combinatur, licebit eius valorem verum ex observationibus supponere, eo fine, ut tanto facilior exactiorque evadat determinatio reliquarum quantitatum. Quodsi vero absoluto hoc determinandi negotio ex primo termino aequationis I. idem valor pro litera α prodeat, quem observationes ostendunt; indicio id erit, theoriam gravitatis Newtonianam, qua vires Solis, Terrae, atque Lunae, quadratis distantiarum reciproce statuuntur propertionales, veritati atque observationibus esse consentaneam; sin minus hoc accidit, calculo ceteroquin accurate peracto, correctione eam indigere docebit'*) (Mayer 1767, p. 33).

[44] Gautier signalled this advantage (Gautier 1817, p. 63–71).

[45] Euler (1753).

But convergence of series is not a trivial matter in the planetary and lunar theories, because small divisors may cause coefficients to blow up in the integrations, as we will see later.

5.4.8 Working Out the Expansions

Integrations

[11] As explained in Sect. 5.4.7, Mayer has to integrate three equations: (5.14), (5.10), and (5.9).

Integration of (5.14) and division by h yields

$$-\frac{P}{h} = \frac{a}{\alpha h} \cos \alpha p + \frac{b}{\beta h} \cos \beta p + \frac{c}{\gamma h} \cos \gamma p + \cdots .$$

There is no need for an integration constant here because P is the integral in equation (5.7) whose constant term has been absorbed in h.

Next, integration of (5.10) yields

$$\varphi = p + \frac{a}{\alpha^2 h} \sin \alpha p + \frac{b}{\beta^2 h} \sin \beta p + \frac{c}{\gamma^2 h} \sin \gamma p + \cdots . \qquad (5.15)$$

An integration constant might have been expected here. Its absence from Mayer's theory implies that $\varphi = 0$ when $p = q = t = 0$. This condition is extremely desirable: a non-zero integration constant would mean that $\varphi = 0$ at an instant $p = p_0 \neq 0$, and in that case either p_0 or the integration constant would necessarily depend on the number of periodic terms that are considered in the series approximation of (5.15).[46]

To integrate (5.9), Mayer first put

$$z = 1 - x = A \cos \alpha p + B \cos \beta p + C \cos \gamma p + \cdots ,$$

so, provided that $z < 1$ (this is a realistic condition because the earth–moon distance is known to be finite over a very long period):

$$x^{-2} = (1 - z)^{-2} = 1 + 2z + 3z^2 + 4z^3 + 5z^4 + \cdots .$$

Termwise backsubstitution and reordering, keeping of the fourth order only the terms involving A^4, gives[47]

---

[46] See also the earlier remarks on integration constants on pp. 70 and 72, and Godfray (1852, p. 30).

[47] An expression such as $\cos(\alpha + \beta)p$ means $\cos((\alpha + \beta)p)$, and $a \cos(\alpha \pm \beta)p$ indicates that there are two such terms: $a \cos((\alpha + \beta)p)$ and $a \cos((\alpha - \beta)p)$.

$$x^{-2} = 1 + \tfrac{15}{8}A^4$$

$$+ \tfrac{3}{2}A^2 + \left(2A + 3A^3 + 6AB^2 + 6AC^2\right)\cos\alpha p + \left(\tfrac{3}{2}A^2 + \tfrac{5}{2}A^4\right)\cos 2\alpha p$$

$$+ \tfrac{3}{2}B^2 + \left(2B + 3B^3 + 6BA^2 + 6BC^2\right)\cos\beta p + \tfrac{3}{2}B^2\cos 2\beta p$$

$$+ \tfrac{3}{2}C^2 + \left(2C + 3C^3 + 6CA^2 + 6CB^2\right)\cos\gamma p + \tfrac{3}{2}C^2\cos 2\gamma p$$

$$+ 3AB\cos(\alpha\pm\beta)p + 3AC\cos(\alpha\pm\gamma)p + 3BC\cos(\beta\pm\gamma)p$$

$$+ A^3\cos 3\alpha p + 3A^2B\cos(2\alpha\pm\beta)p + 3A^2C\cos(2\alpha\pm\gamma)p$$

$$+ B^3\cos 3\beta p + 3B^2A\cos(2\beta\pm\alpha)p + 3B^2C\cos(2\beta\pm\gamma)p$$

$$+ C^3\cos 3\gamma p + 3C^2A\cos(2\gamma\pm\alpha)p + 3C^2B\cos(2\gamma\pm\beta)p$$

$$+ 6ABC\cos(\alpha\pm\beta\pm\gamma)p + \tfrac{5}{8}A^4\cos 4\alpha p + \cdots .$$

$$(5.16)$$

Mayer said that he kept the fourth order in A only as a check on the accuracy of his results, in the following sense. It is evident from experience that the lunar motion is nearly elliptical (the sun causing only small perturbations), hence A is the dominant coefficient in the series (5.13). The terms involving $A^4 \approx (1/20)^4$, which are of the fourth order, should be negligible; Mayer keeps them in order to check if this is really true. If so, then he expects that other terms are justly neglected, because of the dominance of A.[48]

Expression (5.16) is useful in two ways. Mayer uses the constant part of it to fix h: considering that *on average* $dq/dp = 1$, it follows from (5.9) that

$$h = 1 + \tfrac{3}{2}A^2 + \tfrac{3}{2}B^2 + \tfrac{3}{2}C^2 + \tfrac{15}{8}A^4, \qquad (5.17)$$

to the fourth order in A. This expression serves him later when he needs a numerical value for h: he will then substitute values for A, B, C, ... derived from prior results. Second, he substitutes the variable part of (5.16) in (5.9) whereafter termwise integration yields

---

[48] Mayer expressed this a bit further on in his tract as follows: 'In order that yet no scruples arise with this omission, we will everywhere retain the largest of those terms that are rejected. For, after this has been done, if the terms that flow on and on from this source into the longitude of the moon are found to be very small, then it will be safe to conclude from that fact, that the remaining terms too, which might have come up from the omitted ones, are of no importance ('*Ne tamen ullus scrupulus propter hanc obmissionem oriatur, retinebimus ubique maximum eorum terminorum, qui rejiciuntur. Quodsi enim, re peracta, termini, qui hinc in longitudinem Lunae redundant, minimi inveniantur, tutissime inde colligi poterit, neque reliquos, qui ex obmissis oriri potuissent, alicuius momenti esse*') (Mayer 1767, pp. 12–13).

$$q = p + \frac{2A + 3A^3 + 6AB^2 + 6AC^2}{\alpha h} \sin \alpha p + \frac{3A^2 + 5A^4}{4\alpha h} \sin 2\alpha p$$

$$+ \frac{2B + 3B^3 + 6BA^2 + 6BC^2}{\beta h} \sin \beta p + \frac{3B^2}{4\beta h} \sin 2\beta p$$

$$+ \frac{2C + 3C^3 + 6CA^2 + 6CB^2}{\gamma h} \sin \gamma p + \frac{3C^2}{4\gamma h} \sin 2\gamma p$$

$$+ \frac{3AB}{\alpha \pm \beta} \sin(\alpha \pm \beta)p + \frac{3AC}{\alpha \pm \gamma h} \sin(\alpha \pm \gamma)p + \frac{3BC}{\beta \pm \gamma} \sin(\beta \pm \gamma)p$$

$$+ \frac{A^3}{3\alpha h} \sin 3\alpha p + \frac{3A^2 B}{(2\alpha \pm \beta)h} \sin(2\alpha \pm \beta)p + \frac{3A^2 C}{(2\alpha \pm \gamma)h} \sin(2\alpha \pm \gamma)p$$

$$+ \frac{B^3}{3\beta h} \sin 3\beta p + \frac{3B^2 A}{(2\beta \pm \alpha)h} \sin(2\beta \pm \alpha)p + \frac{3B^2 C}{(2\beta \pm \gamma)h} \sin(2\beta \pm \gamma)p$$

$$+ \frac{C^3}{3\gamma h} \sin 3\gamma p + \frac{3C^2 A}{(2\gamma \pm \alpha)h} \sin(2\gamma \pm \alpha)p + \frac{3C^2 B}{(2\gamma \pm \beta)h} \sin(2\gamma \pm \beta)p$$

$$+ \frac{6ABC}{(\alpha \pm \beta \pm \gamma)h} \sin(\alpha \pm \beta \pm \gamma)p + \frac{5A^4}{32\alpha h} \sin 4\alpha p + \cdots . \tag{5.18}$$

Mayer has to invert this series further on in the theory, in order to free the solution from p and express it in terms of the mean motion, q. He cannot work around the series inversion by integrating $dp/dq = hx^2$, because x is expressed as a function of the independent variable p.[49]

Evolution of Terms $1/(x^i y^j)$, $\cos(j\omega)$, $\sin(j\omega)$, etc.

We turn our attention to (I′) and (II′), with X and Y as in the right-hand sides of (I) and (II). In particular, we note those terms in them that have not yet been expressed as a function of the independent variable p. They will be expressed as such, in series, now.

Mayer develops the factors x^{-j} ($3 \le j \le 5$) in the same manner as our example x^{-2} above, whereafter he simplifies the end result slightly by putting $1 + 3(A^2 + B^2 + C^2 + \cdots) \approx h^2$. In order to work to fourth order, Mayer needs to keep only the terms of second order in the indeterminates A, a, ... when expanding terms that become multiplied by the small factor n^2 (which is of second order). As a check on accuracy,[50] he keeps the largest of the third-order terms.

To develop y^{-3} (and similarly y^{-4}) he proceeds as follows. He assumes (reasonably) that the apparent solar motion is undisturbed by the gravitational pull of the moon, so that the apparent solar orbit around the earth is

---

[49] See also footnote 36.

[50] See footnote 48 and the text leading up to it.

elliptical. Let σ be the true solar anomaly and let ϵ be the eccentricity of the solar orbit. Then

$$y^{-1} = \frac{1 - \epsilon \cos \sigma}{1 - \epsilon^2} = 1 + \epsilon^2 - \epsilon \cos \sigma + \mathcal{O}(\epsilon^3),$$

therefore

$$y^{-3} = 1 + \tfrac{9}{2}\epsilon^2 - 3\epsilon \cos \sigma + \tfrac{3}{2}\epsilon^2 \cos 2\sigma + \mathcal{O}(\epsilon^3).$$

Approximation to the second order in ϵ is sufficiently accurate when working to fourth order, because $\epsilon \approx 0.0168 \approx 1/60$ is a quantity of the first order, and all terms involving y^{-3} get multiplied by a factor n^2, which is of the second order.

Next, Mayer expresses σ in the mean solar anomaly $\varsigma = nq$ using the solar equAtion of centre, disregarding the very slight motion of the solar apogee:

$$\sigma = nq - 2\epsilon \sin nq + \tfrac{5}{4}\epsilon^2 \sin 2nq + \mathcal{O}(\epsilon^3),$$

and hence

$$\cos \sigma = \epsilon + \cos nq - \epsilon \cos 2nq + \mathcal{O}(\epsilon^2),$$
$$\cos 2\sigma = \cos 2nq + \mathcal{O}(\epsilon).$$

Using only the largest terms of (5.18) to express the arguments of the cosine terms as functions of p, Mayer arrives at

$$y^{-3} = 1 + \tfrac{3}{2}\epsilon^2 - 3\epsilon \cos np + \tfrac{9}{2}\epsilon^2 \cos 2np$$
$$\mp \frac{3An\epsilon}{\alpha} \cos(\alpha \pm n)p \mp \frac{3Bn\epsilon}{\beta} \cos(\beta \pm n)p \mp \frac{3Cn\epsilon}{\gamma} \cos(\gamma \pm n)p.$$

By similar procedures, Mayer finds expansions first for ω, which denotes the difference of longitudes of the true moon and true sun, and then for $\sin \omega$ and other trigonometric functions of ω. These expansions are expedited by the definition of $r := 1 - n$, so that $rq = q - nq$ is the difference of longitudes of the mean moon and the mean sun. For instance, $\sin \omega$ expands in

$$\sin \omega = \sin rp + \epsilon \sin(r + n)p - \epsilon \sin(r - n)p$$
$$\pm \tfrac{1}{2}\left(\frac{a}{\alpha^2 h} - \frac{2An - 3A^3 n}{\alpha h} \right) \sin(r \pm \alpha)p \pm \frac{3A^2 n}{8\alpha} \sin(r \pm 2\alpha)p$$
$$\pm \tfrac{1}{2}\left(\frac{b}{\beta^2 h} - \frac{2Bn}{\beta h} \right) \sin(r \pm \beta)p \cdots .$$

Finally, Mayer has to face the fact that X in (I$'$) depends crucially on the unknown lunar latitude l, through the term $gh^{-2} \cos^3 l$. There is no chance to solve the latitude equation (III$'$) first because it depends on x being solved. He gets out of the snag by assuming a previously obtained[51] solution for l.

---

[51] '...from my earlier computations' ('*ex calculis meis prioribus*') (Mayer 1767, p. 22). Mayer may here be addressing computations made in the winter of 1753–1754, of which he

This brings in another period-related number i, which denotes the ratio of the lengths of the draconic and tropical months, such that iq is the distance of the mean moon from the mean ascending node.

5.4.9 Filling in the Numbers

[12] We now come to the coefficients B, b, C, c, $\ldots$ in the series (5.13) and (5.14). To compute their values, Mayer substitutes the expansions that he has found so far into (I′) and (II′), in order that every term therein is expressed as a function of his independent variable p. After he has collected terms with common arguments together, two lengthy formulae remain, having 122 and 68 terms respectively: one of these will provide conditions on B, C, $\ldots$, the other on b, c, $\ldots$. For convenience I reproduce them here in a symbolic and much shorter form:

$$\sum_{j=1}^{122} K_j \cos \lambda_j p = 0, \tag{I″}$$

$$\sum_{j=1}^{68} L_j \sin \lambda_j p = 0. \tag{II″}$$

In this modern notation, the λ_j stand for expressions such as $r+2\alpha$ and $\beta-n$, related to the periods of the perturbations of lunar motion. They involve the known constants α, n, $r = 1 - n$, and i, and the indeterminates β, γ of which we are to speak shortly. The factors K_j and L_j stand for combinations of the constants h, k, n, ϵ, A, α, which are known from observations, and the unknowns a, B, b, β, C, c, γ. By way of illustration, the factors for $j = 10$ are:

$$K_{10} = \tfrac{3}{2}n^2 \left(\left(\frac{2Cn}{\gamma h} - \frac{c}{\gamma^2 h} \right)(1 + \tfrac{3}{2}\epsilon^2) + \frac{3C\tau}{2h^2}(1 - \tfrac{5}{2}\epsilon^2) \right), \tag{5.19}$$

$$L_{10} = \tfrac{3}{2}n^2 \left(\left(\frac{2Cn}{\gamma h} - \frac{c}{\gamma^2 h} \right)(1 + \tfrac{3}{2}\epsilon^2)h^{7/3} + \frac{2C\tau}{h}(1 - \tfrac{5}{2}\epsilon^2) \right), \text{ and} \tag{5.20}$$

$$\lambda_{10} = 2r - \gamma,$$

where $\tau \approx 1$ is Mayer's shorthand for a specific complicated expression involving all the unknown constants.

Mayer is silent in his theory about how to determine the coefficients. He merely asserts that it would be a tedious and prolix job to explain, but that his

gave an extensive report in a letter to Euler on 6 March 1754 (Forbes 1971a, pp. 81–83). His original computations of that period are in Cod. μ_4, fol. 6 onwards. These computations were clearly extended during the preparation of *Theoria Lunae*. Fol. 21 of the same manuscript contains nearly the same expansion for $\cos^3 l$ as is found in *Theoria Lunae*.

Display 5.3 The first 12 expressions for K_j, L_j, and λ_j in simplified form

j	K_j	L_j	λ_j
1	$+A(\alpha^2 - (f')^2) + 2a/\alpha h - 86g$	$-a$	α
2	$+B(\beta^2 - f^2) + 2b/\beta h$	$-b$	β
3	$+C(\gamma^2 - f^2) + 2c/\gamma h$	$-c$	γ
4	$+e_5 - 3516g$	$+e_5 h^{7/3}$	$2r$
5	$+t_1 + (3/2A + 15/4A^3)e_5/h^2 + 36g$	$+t_1 h^{7/3} + (2A + 15/2A^3)e_5/h$	$2r + \alpha$
6	$-t_1 + (3/2A + 15/4A^3)e_5/h^2 - 126g$	$-t_1 h^{7/3} + (2A + 15/2A^3)e_5/h$	$2r - \alpha$
7	$+t_2 + 3e_5/2h^2 B$	$+t_2 h^{7/3} + 2e_5/hB$	$2r + \beta$
8	$-t_2 + 3e_5/2h^2 B$	$-t_2 h^{7/3} + 2e_5/hB$	$2r - \beta$
9	$+t_3 + 3e_5/2h^2 C$	$+t_3 h^{7/3} + 2e_5/hC$	$2r + \gamma$
10	$-t_3 + 3e_5/2h^2 C$	$-t_3 h^{7/3} + 2e_5/hC$	$2r - \gamma$
11	$+e_2 - e_4 - 72g$	$(+e_2 - e_4)h^{7/3}$	$2r + n$
12	$-e_2 - e_4 + 186g$	$(-e_2 - e_4)h^{7/3}$	$2r - n$
13	$\ldots$	$\ldots$	$2r + 2\alpha$

method contains nothing that is not already known to anyone who is versed in the modern methods of analysis.[52] His remark is somewhat comparable to Clairaut's *Avertissement* occurring in logically the same place of his lunar theory.[53] Gautier pitied that Mayer did not describe more fully how he worked out the coefficients, and subsequently conjectured how it might have been done.[54] I have been able to glean Mayer's procedure, to be discussed next, from his manuscripts;[55] Gautier's reconstruction is generally in accord with it but considerably less detailed.

Display 5.3 and some examples will help to gain an understanding of the principle of Mayer's procedure. The display lists simplified expressions for λ_j, K_j, and L_j for $j = 1, \ldots, 12$ in (I″) and (II″). These expressions have been simplified in order to make it easier to see how they serve to determine the unknowns B, C, b, c. All the factors of lesser importance have to a large extent been gathered together in the subscripted letters e_i and t_i, which do not appear in Mayer's work. The definitions of e_i and t_i are unimportant here; to get the idea, one may compare (5.19) and (5.20) with their simplified forms K_{10} and L_{10} in Display 5.3.

As far as possible Mayer substitutes numbers, as listed in Display 5.1, for the constants occurring in (I″) and (II″). These numbers are deduced from observations. In part, they may be interpreted as the boundary conditions to the differential equations, such as the eccentricities of the orbits (A, ϵ) and the mean distance ratio of sun and moon (k). Additionally, Mayer uses his

---

[52] Mayer (1767, p. 37).

[53] Clairaut (1752b, p. 53).

[54] Gautier (1817, pp. 70–72).

[55] Cod. μ_4, fol. 14v onwards.

'prior results' to estimate the composite quantities h, τ, f^2, e_i, and t_i, which depend in part on the very coefficients B, b, etc. which are to be determined.

The use of these prior results invalidates, in principle, the independence of his theory, because he inserts data that the theory is supposed to yield. Mayer argues that the end result will be sufficiently accurate with these data put in; e.g., h, τ, and f^2 are close to unity, making their exact values less critical for the theory, since they appear as factors. Perhaps his practice could be formally justified in an iterative procedure.

Then, with numbers substituted, all quantities in Display 5.3 are known numerically except B, b, β, C, c, γ; in addition α is considered known except when $j = 1$ (cf. p. 78). In particular, every λ_j which is free of β and γ is known, again exempting $\lambda_1 = \alpha$. Now look at the right-hand sides of (I'') and (II''): they are both nil. For all practical purposes, this implies conditions on the coefficients:

$$\sum_{j \in J} K_j = 0, \quad \sum_{j \in J} L_j = 0 \qquad (5.21)$$

whenever J is a maximal set satisfying that $\lambda_i = \lambda_j$ for all $i, j \in J$. These will hereafter be referred to as the vanishing conditions.

Mayer's key idea is to assign to β or γ consecutively every one of the known values $2r$, $2r \pm \alpha$, $2r \pm n$, $\ldots$, in other words, precisely those values of λ_j which do not contain β and γ. For each value assigned to, say, β, there is an individual term $K_2 \cos \lambda_2 p$, i.e.,

$$\left(B(\beta^2 - f^2) - 2b/\beta h\right) \cos \beta p$$

in (I'') and an individual term $L_2 \sin \lambda_2 p$, i.e.,

$$-b \sin \beta p$$

in (II''). Note that the only unknowns in the expressions for K_2 and L_2 are B and b; their values can consequently be determined with the help of the vanishing conditions (5.21). When their values are known for all consecutive assignments to β and γ, substituting them in (5.13) and (5.15) will yield the position of the moon.

Mayer now has a price to pay for the efficiency of his notation which we signalled before, namely that there will be many different instances of β, γ and B, b, C, c. For this reason we will append subscript indices where appropriate. In his private papers, Mayer sporadically employed a superscript notation with the same effect.

We will now illustrate the procedure with an example for the calculation of the 11th terms in (I'') and (II''). We see in Display 5.3 that $\lambda_{11} = 2r + n$, therefore we assign $\beta = 2r+n$. In order to find values for $B_{(2r+n)}$ and $b_{(2r+n)}$, we consider the vanishing condition:

$$(K_2 + K_{11}) \cos(2r + n)p = 0 \quad \text{for all } p.$$

With K_2 and K_{11} as in Display 5.3, the required condition is that

$$B_{(2r+n)}\big((2r+n)^2 - f^2\big) + \frac{2b_{(2r+n)}}{(2r+n)h} + e_2 - e_4 - 72g = 0$$

or equivalently

$$B_{(2r+n)} = \frac{e_2 - e_4 - 72g + 2b_{(2r+n)}/((2r+n)h)}{f^2 - (2r+n)^2}.$$

The expression on the right-hand side cannot be computed, because the coefficient $b_{(2r+n)}$ is not yet known. In this particular example, the latter can be easily computed using the vanishing condition for L_{11} and L_2, but in other cases Mayer resolves the issue by filling in its value guided by previous experience. He borrowed its value probably from a lunar table coefficient which he had previously corrected to observations. In specific cases, to be dealt with below, drawing upon such previous experience provides a considerable simplification of the computations.

With the right-hand side established, a value for coefficient $B_{(2r+n)}$ rolls out, and the term $B_{(2r+n)} \cos(2r + n)p$ is required in the series (5.13) in order to cancel in (I'') the term $(e_2 - e_4 - 72g) \cos(2r + n)p$ produced by the gravitational accelerations.

Similarly, to find coefficients $B_{(2r-n)}$ and $b_{(2r-n)}$, the process is repeated with $\beta = 2r - n$, to yield terms that cancel against $(-e_2 - e_4 + 186g) \cos(2r - n)p$ and $(-e_2 - e_4)h^{7/3} \sin(2r - n)p$; then it is repeated again for $\beta = 2r + 2\alpha$ and other vaules of β, until all terms have been canceled.

5.4.10 Odds and Ends

With the basic principle of coefficient determination now mastered, we turn to some of the details. To begin with, we have used the specific choice of $\beta = 2r + n$ to cancel exactly one term in (I'') and one in (II''). Yet β occurs in many more terms in these series (such as the 7th and 8th) and we have to give due consideration to those occurrences as well. For instance, this particular choice of β renders $\lambda_7 = 4r + n$, and the seventh terms with arguments $(4r + n)p$ have to be annihilated by a similar calculation as in the example above. This entails first the calculation of the coefficients K_7 and L_7 with $B = B_{(2r+n)}$ and $b = b_{(2r+n)}$, and then the calculation of $B_{(4r+n)}$ and $b_{(4r+n)}$ in K_2 and L_2 using $\beta = 4r + n$ and the vanishing condition. In most instances these contributions are luckily negligible and of no concern, because the coefficients $B_{(2r+n)}$ and $b_{(2r+n)}$ which enter the expressions for K_7 and L_7 are usually small.

However, in some instances the contribution *is* significant. This happens when β, γ are assigned values of $2r$ and $2r - \alpha$, associated with the largest perturbations of lunar motion: evection and variation, which come with large coefficients.[56] Those values $2r$ and $2r - \alpha$ are so important that Mayer goes once through the series (I'') and (II'') with $\beta = 2r$ and $\gamma = 2r - \alpha$ simultanuously. The single reason to include β- and γ-terms in the series was to compute the terms with $\lambda_j = 2r \pm \beta \pm \gamma$ or $\lambda_j = \alpha \pm \beta \pm \gamma$, which are sizeable precisely in the case just mentioned.

An other point of interest is that the assignment $\beta = 2r$ makes $\lambda_8 = 2r - \beta = 0$, and hence $\cos \lambda_8 p = 1$, $\sin \lambda_8 p = 0$, for all p. Mayer seems to disregard these constant terms silently. The rationale, as I see it, comes in three parts. In the first place, if $\lambda_j = 0$ then $K_j \sin \lambda_j p = 0$ for all p, therefore such a term makes no contribution to lunar longitude. Second, at the same time $L_j \cos \lambda_j p = L_j$, which would give a constant contribution to the radial distance; yet this contribution must have been included in the mean distance taken from observations, therefore it is safe to disregard this term too. Third, the computation of the coefficients B, b, ... will yield values that may prove inaccurate, especially when the computations involve small divisors. These occur when either $\beta \approx 0$ (giving bad B and b) or $\beta^2 \approx f^2$ (giving bad B only). The former condition applies when $\beta = 2(\alpha - i) \approx -0.03$, the latter when $\beta = r + n = 1$ and $\beta = 2i - \alpha \approx 1.02$. Their coefficients are best determined from observations.[57]

To make a third remark, we return to the motion of the apsidal line. Mayer computes the value of α from the factor K_1 in basically the same way as outlined in Sect. 5.4.9, except that the roles of A and α are reversed: A is known and α is the unknown quantity. Mayer finds $\alpha = 0.9915965$,[58] which is in excellent agreement to the observational value quoted in Display 5.1. He computes a lunar apogee advance of $6'38.6''$ per day, which compares favourably to the value from observations of $6'41''$ per day. He had apparently kept the right number of higher order terms to include the notorious half of the apogee movement which had plagued Newton, Clairaut, and others so much. This might perhaps be another of the lessons that he learnt from Clairaut, although some doubt arises because Mayer neither mentioned that the success is due to those higher order terms, nor did he write them down in his theory.

---

[56] The arguments $2rq$ and $(2r - \alpha)q$ are 2ω and $2\omega - p$ in the standard notation of Sect. 1.3.

[57] In the paragraph devoted to the adjustment of theory to observations, Mayer remarks that 'many terms appear [...] which the theory, even if treated with the utmost care, cannot furnish accurately, for reasons well known to anyone who has exercised his vigour and patience in this matter' (Mayer 1767, p. 50); also quoted below in fn. 2 in Sect. 8.1.1. This problem of small divisors was first signalled by Euler in his treatise on the great inequality of Jupiter and Saturn, cf. Wilson (1985, p. 105).

[58] Mayer (1767, p. 41).

5.5 Solved Equations

[13] With the coefficients in (I′) and (II′) determined, Mayer is able to express x, q, and φ as trigonometric series in the independent variable p. We may consider them as functions of p, thus: $x = x(p)$, $q = q(p)$, and $\varphi = \varphi(p)$. By an inversion of series he then expresses $p = p(q)$ as a series in q, whereupon both $x = x(p(q))$ and $\varphi = \varphi(p(q))$ may be expressed as series in q. Clairaut, in his lunar theory, explained in considerable detail how to invert series.[59] The lengthy and not very inspiring computations result in expressions of the reciprocal earth–moon distance x and of the lunar true longitude φ in terms of the lunar mean longitude q. Together with the separately obtained solution of the latitude equation, Mayer then has expressions for each of the three coordinates of the moon's position. It is what we call a single-stepped solution of the problem of lunar motion, meaning that no intermediate changes are made to the arguments of the equations: every single term in the solution is the sine or cosine of a mean motion argument. Mayer's single-stepped solution for the longitude coordinate is reprinted in the left-hand column of Display 7.1 in Chap. 7. In that chapter, we will see how this single-stepped solution relates to the multistepped format of the tables. In Chap. 8 we turn to the fitting of coefficients to observations, which was only briefly mentioned in *Theoria Lunae*.

5.6 Concluding Remarks

Similar to his contemporaries Euler, Clairaut, and d'Alembert, Mayer developed an attack on the 'main problem' of the moon's motion from dynamical principles and in spherical coordinates. The characteristic features of Mayer's lunar theory are:

- The choice of independent variable, by which the derivatives of the coordinates are each expressed in their own differential equation
- A way to manage the multitude of terms in the trigonometric series, by which he prevents getting impaired by them
- One equation for latitude, instead of separate equations for node and inclination

Mayer was probably the first to treat the latitude equation in a way similar to the treatment of the longitude equation.[60] At the same time, his deprecating the separation of latitude into node and inclination entailed a disregard of the method of variation of constants with which Euler was experimenting at the time, an approach that proved to be very successful later in the century

---

[59] There are three lemmata and a problem to that extent in Clairaut (1752b, pp. 55–59).

[60] Cf. Gaythorpe (1957, p. 143) and Godfray (1852, p. iv).

and that is regarded as fundamental today. His disregard makes it even less likely that he had an osculating orbit in mind with the osculating anomaly as independent variable.

In some respects, Mayer's lunar theory is not completely self-supporting. He assumed the mean motion of the apogee unknown in some places, but known (for simplicity) in other places. Likewise, he took knowledge of the lunar latitude for granted in solving the differential equation for lunar longitude. He also used previously obtained magnitudes of many equAtions when he had to find the coefficients of the series numerically. So, even before the adjustment of theory to observations, Mayer took much more from observationally obtained knowledge than the necessary minimum of six integration constants plus the ratios of masses and mean distances of sun and moon. But whenever Mayer reverted to assuming values based on observations or previous results, he made an effort to explain that an approximate value would suffice, so that his theory would not depend crucially on it.[61]

Why was Mayer's theory more accurate than other theories? In the introduction to *Theoria Lunae*, Mayer advertized that the coefficients in his theory differed little from those in his tables. The latter, being adjusted to observations, were much more accurate than those of his rivals Euler, Clairaut, and d'Alembert. Mayer's theory, up to the point where the coefficients are actually computed, was not of a higher order of accuracy than the contemporary theories (they were all of the second order in the lunar eccentricity), nor did Mayer include physical causes that were neglected by others. Did Mayer have better values for the fundamental orbital constants? The four constants with a fundamental role in Clairaut's theory are: the lunar and solar eccentricities, the ratio of lunar and solar mean motions, and the ratio of their parallaxes. Of these, only Clairaut's value of the latter differs markedly from Mayer's value: Mayer supposed the solar parallax $10.8''$, Clairaut $12''$.[62] It seems unlikely that the differences between Mayer's and Clairaut's final coefficients (and, similarly, those of other theories) are due mainly to the small differences in such fundamental constants. Probably the only reason why Mayer's theory was so much more accurate is its greater dependence on empirical data, which went directly into the coefficients. One wonders then whether the difference between his theory and fitted tables might be so slight *because* the coefficients of the theory depend so much on empirical results.

When we compared Mayer's *Theoria Lunae* with Euler's *Theoria Motus Lunae*, we found that the two run largely in parallel up to the formulation of the differential equations and the forces. From there on the two theories start to deviate: Euler chooses true anomaly as his independent variable, but Mayer chooses an independent variable which somewhat resembles osculating true anomaly. Although both start from the same basic equations and share some of the same techniques, such as the application of trigonometric series,

---

[61] Examples are in §§35, 39, 41, 42 of *Theoria Lunae*.

[62] Clairaut (1752b, p. 55).

the theories proceed differently: they deviate in the ways that the game of approximation is played. Euler considered various physical effects represented in the equations one by one, which may provide a better understanding of the nature of the various perturbations. Mayer had an efficient way to deal with all effects at once but he needed to feed in many more a priori values for constants. Unlike Euler, Mayer saw no reason to divide the latitude equation into separate differential equations of the first order for node and inclination. This means that Mayer did not appreciate Euler's approach that went in the direction of the method of variation of constants.

It is harder to say to what extent *Theoria Lunae* was influenced by Clairaut's *Theorie de la Lune*. Mayer had decided to model the beginning of his theory after Euler's, thereby leaving less occasion for an explicit influence of Clairaut. Any extant similarities are further obscured because Euler himself had drawn inspiration from Clairaut's work. Yet all the three theories employ a different independent variable. The use of an integrating factor to integrate (II), and the ensuing substitution for the integral on its right-hand side, were common in all, as was the application of trigonometric series. Those were the tricks of the trade. Clairaut had possibly two direct influences on Mayer: the method of computation of the coefficients, as outlined in Sect. 5.4.9, and the inversion of series: both were treated to some extent in Clairaut's theory. Conceivably, Clairaut also inspired Mayer's choice of independent variable.

Chapter 6
The Horrocks Legacy

During 1752, a major change took place in Mayer's thinking about modelling of the moon's motion, which affected the procedure to be followed when computing a lunar position from the tables. Initially the tables were applied in the single-stepped fashion (using the terminology explained in Chap. 4), but during 1752 Mayer switched to a multistepped procedure. He would basically adhere to the multistepped procedure ever after, and change only minor details later on.

The purpose of the present chapter is to investigate the origin of this multistepped procedure. We will see that Mayer developed it out of a lunar theory of Newton's, first printed in 1702, which at best has a very troublesome relation to the theory of gravitation. This dependence of Mayer's tables on Newton's 1702 lunar theory has never been noted before. On the contrary, it is usually said that Mayer's tables were in some way based on Euler's lunar theory, and that their advent made those founded on Newton's 1702 lunar theory obsolete.[1] It will here be shown that Newton's theory exercised, through Mayer's tables, a much more profound impact on eighteenth century positional astronomy than has hitherto been thought.

We will see in the next chapter how to transform any multistepped scheme into a single-stepped scheme that is equivalent to it for all practical purposes. This implies that none of the many possible schemes has a real advantage over any other as far as achievable accuracy is concerned. Nonetheless, we will see that Mayer attained a most striking improvement in accuracy as soon as he adopted the multistep procedure. The manuscripts that witness the emergence of his new scheme suggest that he was at that time unable to provide a complete and coherent lunar theory to back up his new tables. The

---

[1] The relation between Mayer's and Euler's lunar theories has been discussed in the previous chapter. Statements effectively equivalent to 'Newton out, Mayer in' can be found in, for example, Wilson and Taton (1989, p. 162), Kollerstrom (2000, pp. 233–234), Whiteside (1976, p. 324). Petrus Frisius came very close to recognizing Mayer's dependence on Newton's lunar theory, only to conclude that Mayer had fitted Euler's theory to observations (Frisius 1768, pp. 272–273, 357).

S.A. Wepster, *Between Theory and Observations*, Sources and Studies in the History of Mathematics and Physical Sciences, DOI 10.1007/978-1-4419-1314-2_6,
© Springer Science+Business Media, LLC 2010

development of the multistepped scheme happened apparently for pragmatic rather than theoretical reasons, and it is highly unlikely that Mayer had a valid theoretical justification for it over the single-step procedure.

The influence of Newton's lunar theory on Mayer provides yet another argument against the commonly held belief that Mayer's tables are merely Euler's equations (either in the astronomical or the mathematical sense) with their coefficients adjusted to observations: such a belief leaves the multistep procedure unexplained. We also see the role of the calculus in the development of Mayer's 1753 kil tables reduced to a subsidiary one, at most. But a final judgement of the relative merits of the calculus, Newton's theory, and observations is more delicate and will be postponed until the final chapter.

The current chapter is organized as follows. In Sect. 6.1 we will start with a study of the 1702 'theory' of Isaac Newton, which is, as we will see, rather a set of rules to construct lunar tables than a theory in the modern sense of the word. Those rules are presented in Sect. 6.2. Then follow two sections on a crucial ingredient of Newton's lunar theory that had developed out of an idea of Jeremiah Horrocks (1618–1641): a variable eccentricity and direction of the apsidal line of the lunar orbit. The kinematics of Horrocks's idea is explained in Sect. 6.3, and in Sect. 6.4 it is translated into the language of trigonometric functions. The results will go slightly beyond what has already been published on the subject.

Section 6.5 is devoted to the lunar tables of Pierre Charles Lemonnier. His tables were probably the widest available implementation of Newton's prescriptions, and they were instrumental in the transmission of Newton's rules to Mayer. Mayer's assimilation of the Newtonian Lemonnier tables is the subject of Sect. 6.6, where we study the relevant manuscripts. It will become clear that Mayer's interest in Newton's theory arose when he was apparently in doubt of how to proceed further in lunar theory.

Then follows an investigation of the measure of success of some of Mayer's table versions. In particular, we compare the accuracy of his tables before and after the multistep reform. The accuracy of Lemonnier's tables is included too. We will discover that the tables based on a multistepped scheme result in higher accuracy than Mayer's initial single-stepped scheme. Some insight will also be gained in the further improvements that Mayer attained. Finally, we will take a look at Mayer's own preface to his printed tables.

6.1 Newton on Lunar Motion, 1702

In 1702, Isaac Newton's *Theory of the Moon's Motion* (*NTM*) appeared, a pamphlet containing a set of rules for the production of tables for the computation of the position of the moon. Four nearly identical editions, three in English and one in Latin, have appeared of that text; the Latin version was first published in David Gregory's *Astronomiae Physicae & Geometricae*

Elementa, 1702. All four have been reproduced in facsimile with a general introduction by I. Bernard Cohen.[2]

NTM is not a theory as one might perhaps expect. In fact, quite different from the modern scientific idea of a theory, *lunar theory* in that time denoted, in the words of Francis Baily, 'rules or formulae for constructing diagrams and tables that would represent the celestial motions and observations with accuracy'.[3] Nothing else could have been expected before the formulation of a causal physical theory of movement of celestial bodies. Even after he had provided precisely such a physical theory, Newton continued to use the word in its habitual sense.

The first edition of Newton's *Principia*, of 1687, contained a significantly more rudimentary lunar theory than the 1702 pamphlet. Somewhat modified and condensed forms of the *NTM* prescriptions found their way into the second (1713) and third (1726) editions of the *Principia*, where they can be found in the Scholium to Proposition 35, Book III, following a *quantitative* examination (ignoring eccentricity) of the variability of the inclination of the lunar orbit, the motion of the nodes, and the variation. In addition, Proposition 25 of Book III called upon the many corollaries of Proposition 66 of Book I for a *qualitative* explanation of all known lunar equations, and some new ones.[4]

A very characteristic feature of *NTM*, which we will discuss at length in the following two sections, was adapted from an older, kinematic, lunar theory of Jeremiah Horrocks. This suggests that Newton's lunar theory was a mix of his own dynamical research and of Horrocks's kinematic model.[5]

---

[2] Newton (1975). Cohen makes a distinction between (1) the 1702 pamphlet, (2) its contents without reference to a specific edition, and (3) Newton's work on lunar theory in general. The abbreviation *NTM* for Newton's Theory of the Moon is in line with Cohen's indication of the second category. An impression of Newton's work on lunar theory is provided in Whiteside (1976); also see *A guide to Newton's Principia*, in Newton et al. (1999), particularly §8.14, *The Motion of the Moon*, by I. B. Cohen, and §8.15, *Newton and the Problem of the Moon's Motion*, by George E. Smith. Kollerstrom (2000) discusses the procedures of *NTM*, assesses its accuracy using computer simulations, and points to various places where *NTM* was used in the eighteenth century, missing – like every other researcher before him – its influence on Mayer. Cook (2000) contains a well-balanced and illuminating view of Newton's work on lunar motion, providing physical insight while avoiding mathematical detail.

[3] Baily (1835, p. 690), quoted in Newton (1975, p. 3). Contrast this to the modern connotation, as exemplified for instance in the on-line edition of the Merriam-Webster dictionary (http://www.m-w.com), which defines theory as 'a plausible or scientifically acceptable general principle or body of principles offered to explain phenomena'. Indeed, not everything that Newton wrote was Newtonian!

[4] Newton's notorious attempt in Book I Proposition 45 on the mean motion of the apsidal line is of no concern to us now; see Waff (1975) for that.

[5] Here the words 'dynamical' and 'kinematical' are used to distinguish between contexts or theories that do, respectively do not, apply the principle of gravitation and the techniques of the infitesimal calculus.

Surprisingly, gravitation was only mentioned once in *NTM*, and then only in the preface, which was certainly not written by Newton but perhaps by Halley:

> This Irregularity of the Moon's Motion depends (as is now well known, since Mr. Newton hath demonstrated the Law of Universal Gravitation) on the Attraction of the Sun, which perturbs the Motion of the Moon [...]. But this being *now* to be accounted for, and reduced to a Rule; by this Theory such Allowances are made for it, as that the Place of the Planet shall be truly Equated.[6]

The text ascribes the cause of the perturbations to the attraction of the sun, and stresses that it is now time to lay down rules to compute the perturbed motion of the moon, but it avoids to aver that the rules are purely deduced from the law of gravitation. Indeed, the main text of *NTM* supplied these rules, as a true theory in the pre-Newtonian sense of that word.

In *Principia*, however, Newton repeatedly averred that he had obtained all his results from application of the law of gravitation to the Sun–Earth–Moon system, but he omitted his derivations in all but the above-mentioned three cases: variation, nodal movement, and inclination. Michael Nauenberg has pointed to certain manuscripts in the Portsmouth collection where Newton apparently applied a perturbation technique that might have allowed him to obtain the results of *NTM*.[7] But Nauenberg's point of view that Newton indeed had provided a dynamical and gravitational basis for the Horrocksian part of his theory is not accepted by everybody. Probably, though, Newton was able to obtain the form of some equations of lunar motion theoretically, whereafter he adjusted the coefficients to observations; at least, he explained to Flamsteed that this was his procedure. Although some of the equations were new discoveries disclosed by the law of gravitation, Whiteside and Kollerstrom maintain that Newton went back to the Horrocksian model in despair, after failing to account for *all* the lunar equations by gravitation.[8] It is of interest that Newton adapted a similar Horrocks-inspired model in an attempt on the Jupiter–Saturn inequality, as Wilson noted.[9]

Surely, the moon had played a crucial role in Newton's discovery of the law of gravitation; yet the intricacies of lunar motion made his head ache, as he confessed to Machin. Although we regard Newton also as an inventor of the differential and integral calculus, the tools at his disposal were quite different from the tools that were developed later in the eighteenth century. Wilson explained it lucidly thus:

> Newton worked out the motions of celestial bodies while thinking predominantly geometrically, and at every step he had to give full account of the dynamics of the problem. In the eighteenth-century approaches, differential equations were formed

---

[6] Newton (1975, pp. 94–95). Cohen discusses the allusion to Halley on pp. 31–32.

[7] For pointers, see Nauenberg (1998, 2000, 2001).

[8] Whiteside (1976), Kollerstrom (2000).

[9] Wilson (1985, p. 17).

on geometrical and dynamical grounds, whereafter the solution lay in the realm of analysis, having to find successive approximations to an analytical function.[10]

All things taken into account, it is safe to say that the theoretical basis of *NTM* is still unclear today.[11]

6.2 The Equations of Newton's Theory

In *NTM*, Newton first specified the epochs and mean motions of the lunar longitude, apogee, and node, and the solar longitude and apogee, and then proceeded to describe the various equAtions. We will now have a more detailed look at these equAtions, representing them in modern, analytical form.[12] We will need the amount of detail included here in this section to appreciate the impact of *NTM* on Mayer later on in the chapter.

NTM's procedure is of the kind where each equAtion affects the arguments before the next equAtion is computed: it can be regarded as a multistep procedure with only one equAtion per step. *NTM* has seven steps for longitude, some very simple, and some more complicated. Several equAtions are subject to seasonal variations due to the varying distance of the earth–moon system from the sun. The central fourth step embodies both the equAtion of centre and the evection combined via a geometrical construction to be discussed in the next two sections. We will now go through Newton's seven steps one by one.

[1] The first lunar equAtions specified in *NTM* are the annual equAtions to lunar longitude, apogee, and node. Newton stated the maximum values of these equAtions as $+11'49''$, $-20'$, and $+9'30''$, respectively. He specified further that they are proportional to the solar equAtion of centre with argument solar mean anomaly, here denoted by ς. From this proportionality, it follows that the equAtions encompass terms proportional not only to $\sin \varsigma$, but also to $\sin 2\varsigma$.[13] The annual equAtion is the only step in Newton's prescriptions that affects the node. In all the following steps, the equAtions apply only to lunar longitude, except the fourth one which also affects the apogee.

[2] To convey the character of Newton's text, I quote here his prescription for the second equAtion:

---

[10] Wilson (2001, p. 178); also see Wilson (1985, p. 69ff.).

[11] Baily (1835, pp. 139–140), Newton (1975, p. 39), Whiteside (1976), Wilson (1995a, p. 50), Kollerstrom (2000). The source of the anecdote of Newton's headache is a notebook of John Conduitt (Whiteside 1976, p. 324).

[12] The research is based mainly on Cohen's edition of *NTM* (Newton 1975, pp. 91–119), with Kollerstrom (2000) as a secondary source.

[13] I indeed found such terms, which were overlooked by Kollerstrom, in Lemonnier's tables to be discussed below. Higher order terms, proportional to $\sin kp$ for $k \geq 3$, contribute less than an arc-second and are therefore undetectable at the precision of Lemonnier's tables.

There is also an *Equation of the Moon's mean Motion* depending on the Situation of her Apogee in respect of the Sun; which is *greatest* when the Moon's Apogee is in an Octant with the Sun, and is nothing at all when it is in the Quadratures or Syzygys.[14] This Equation, when greatest, and the Sun *in Perigaeo*, is 3′56″. But if the Sun be *in Apogaeo*, it will never be above 3′34″. At other distances of the Sun from the Earth, this Equation, when greatest, is reciprocally as the cube of such Distance. But when the Moon's Apogee is any where but in the *Octants*, this Equation grows less, and is mostly at the same distance between the Earth and Sun, as the Sine of the double Distance of the Moon's Apogee from the next Quadrature or Syzygy, to the Radius.

    This is to be *added* to the Moon's Motion, while her Apogee passes from a Quadrature with the Sun to a Syzygy; but it is to be *subtracted* from it, while the Apogee moves from the Syzygy to the Quadrature.[15]

In other words, the second equation depends on the sine of twice the distance of the lunar apogee from the sun, that is $\sin(2\varpi - 2p)$ in Mayer's notation. Its coefficient, says Newton, varies annually between 3′56″ and 3′34″ reciprocally as the cube of the distance of the sun from the earth. When we express the eccentricity of the earth's orbit by $\epsilon \approx 0.0168$, then the cube of the earth–sun distance reciprocally is very nearly $(1 + \epsilon \cos\varsigma)^{-3} \approx 1 - 3\epsilon \cos\varsigma$. Hence the seasonal fluctuation amounts to 3ϵ or approximately $\frac{1}{20}$ of the coefficient at the mean distance. We take the mean coefficient as the arithmetical mean of the annual extremes, or 3′45″. Hence we deduce that Newton's second equation is very nearly $3′45″(1 - 3\epsilon \cos\varsigma) \sin(2\varpi - 2p)$, that is, $(3′45″ - 11″ \cos\varsigma) \sin(2\varpi - 2p)$.

[3] The third equation, represented analytically, is $47″ \sin(2\varpi - 2\delta)$. Newton's description is in the same vein as the quote above, but simpler, because the seasonal change is too small to warrant mention.

[4] The middle of Newton's steps encompasses a kinematic construction borrowed from Horrocks's lunar theory, periodically modifying both the eccentricity of the lunar orbit and the orientation of its apsidal line before the equation of centre is computed using this modified eccentricity and anomaly. I postpone further discussion to the next section.

[5] Next comes the variation, which, like the second equation, has a seasonal component depending on the earth–sun distance. It is approximated analytically by $(35′32″ - 1′53″ \cos\varsigma) \sin 2\varpi$.

[6] Newton's next equation amounts to $2′10″ \sin(2\varpi + \varsigma - p)$. Originally, Newton had specified this equation in *NTM* with the wrong sign. The second and third *Principia* editions corrected the mistake, increased the coefficient to 2′25″, and gave it a treatment conjointly with the equations of the fourth step, whence Newton called it his Second equation of centre. In the single-stepped computational procedure commonly used today, this equation indeed has a negative coefficient.[16]

---

[14] Syzygy and quadrature: cf. fn. 20 on p. 24.

[15] Newton (1975, pp. 105–106).

[16] See Chap. 7 and in particular Display 7.1 for the effect of the multistepped procedure on the coefficients; see for example, Meeus (1998, p. 339) for modern values of coefficients.

[7] EquAtion seven is equivalent to $\left(2'20'' + 54'' \cos(\omega + \varsigma - p)\right) \sin \omega$. This equAtion was omitted in the third *Principia* edition. Kollerstrom noticed that the sign of the $54''$ annual coefficient is negative in some of the *NTM* versions.[17]

So far for the longitude equAtons of *NTM*; the text continues with equAtions for parallax, latitude, and reduction to the ecliptic, which are of no interest for our current discussion.

In the number of equAtions dealt with, *NTM* surpassed every other lunar theory extant at its time of publication. The annual equAtions to apogee and node, prescribed in *NTM*'s first step, were new inventions of Newton.[18] He introduced four other new equAtions: the 2nd, 3rd, 6th, and 7th. The annual equAtion to longitude, present in step 1, and the variation of step 5, had both been discovered by Tycho Brahe. Ptolemy had modelled evection, and the equAtion of centre (or, at least, computational procedures to express the variable velocity of the moon) had been known throughout antiquity.

Newton's precepts were the best means of computing lunar positions in the first half of the eighteenth century. They were indeed used to construct tables, although it took about thirty years before those were widely available. Flamsteed made tables already in 1702, Halley did so too in about 1720, but although Halley's tables were printed, neither his nor Flamsteed's were ever published. Tables based on *NTM* were also produced by Wright (1732), Leadbetter (1735), and several others. Perhaps the first Newtonian tables to be published were those of Peder Horrebow, 1718, but they lacked a widespread distribution. Lemonnier's handbook *Institutions Astronomiques* was very instrumental in spreading Newton's theory in the form of tables.[19]

6.3 Horrocks's Variable Orbit

Gravitational motion in an elliptic orbit respects the area law, hence it is not uniform. As is well known among astronomers, this brings about the equAtion of centre, a correction of the mean motion whose instantaneous value depends on the anomaly (i.e., the angular distance of the orbiting body from an apside) and the eccentricity. The fourth step in Newton's *NTM* sets up an equAtion of centre for a variable lunar orbit: i.e., subjected to a variable apsidal line orientation and a variable eccentricity. Thus the form of the approximate elliptical orbit of the moon is supposed to change in *NTM*. The

---

[17] Kollerstrom (2000, p. 106).

[18] Newton discussed them in Newton et al. (1999, Bk. III, Scholion after Prop. 35), and Newton (1975, pp. 103–105). Wilson (1989b, p. 265) refers to Newton's *Principia* bk. III Prop. 22, where Newton alludes that 'there are other inequalities not observed by former astronomers' mentioning a.o. the inequAlities that these annual equAtions are to correct.

[19] Cf. Kollerstrom (2000, Chap. 14).

Fig. 6.1 The variable
lunar orbit

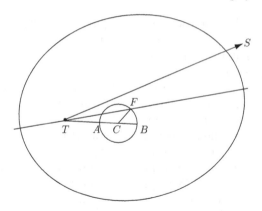

form change is effectuated by imagining that the centre of the elliptical lunar
orbit moves in a small circle about its mean position, as follows.

In Fig. 6.1, let T be the earth, TS the direction of the sun from the earth,
and TC the direction of the lunar apogee corrected by the annual equAtion
of Newton's first step. C is the mean position of the centre of the lunar
orbit. The actual centre is in F, which is taken to revolve on the circle BFA
around C. With respect to the (once equAted) lunar apogee, F revolves twice
as fast as the sun, so $\angle FCB = 2\angle STC$. The length TF between the focus
and centre of the elliptical lunar orbit represents the eccentricity, because the
semi-major axis of the orbit is taken constant. As F revolves around C, the
eccentricity varies between its minimum $TF = TA$ and maximum $TF = TB$.
Furthermore, the apsidal line goes through T and F, and as F rotates, this
line changes its direction so that the apogee rocks back and forth around
its mean position on the line $TACB$ extended. The angle $\angle FTC$ between
the actual and mean apse line is a second equAtion to the apogee position.
The equAtion of centre, being a function of both eccentricity and anomaly,
must then be computed with this modified eccentricity and apogee position.
Figure 6.2 shows three diagrams for three different orientations of the sun
with respect to the mean lunar apsidal line.

The fourth step in NTM provides then (1) a variable eccentricity TF of
the lunar orbit; and (2) a (second) equAtion to the lunar apogee in the angle
FTC; both affecting (3) the equAtion of centre.

Newton provided the following values in his NTM: the eccentricity ranged
from $TA = 0.043319$ to $TB = 0.066782$ (expressed as fractions of the constant
semi-major axis of the orbit); the equAtion of centre would then (as stated
in NTM) attain its extremes of $4°57'56''$ and $7°39'30''$, respectively. These
values imply a mean eccentricity of

$$TC = \tfrac{1}{2}(TA + TB) = 0.0550505$$

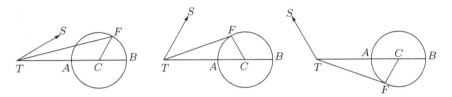

Fig. 6.2 The eccentricity TF and second apse equation angle FTC of the variable orbit in three different states (the lunar orbit itself is not shown)

and radius of
$$FC = \tfrac{1}{2}(TB - TA) = 0.0117315.$$

For future reference, we note the ratio $FC : TC = 0.213104$. The maximum second equation of the apogee is then $\angle FTC = \arcsin(FC/TC) = 12°18'15''$, whereas, surprisingly, Newton in *NTM* asserts the maximum to be $12°15'4''$. In the second edition of *Principia*, all values were revised and brought in agreement to each other, the eccentricity ranging over 0.05505 ± 0.0117275 with a maximum apogee correction of $12°18'$.

Newton included a qualitative description of the variable eccentricity and rocking apsidal line in *Principia* book I, Proposition 66, corollaries 7–9, without explicitly referring to this mechanism of the rotating centre of the moon's orbit. The idea of an ellipse of variable eccentricity and rocking apse had been introduced into the lunar theory by Jeremiah Horrocks, who developed his theory, from 1638 until his early death two years later, from a combination of Keplerian ellipses and Philips Lansbergen's periodical variations of a (circular) orbit. The theory of Horrocks had remained unknown until Wallis published Horrocks's manuscripts in 1672.[20] To this posthumous edition were attached Flamsteed's tables of Horrocks's lunar theory; and it was Flamsteed who later, in 1694, convinced Newton of the quality of Horrocks's theory. Such happened perhaps at a time when Newton's own researches on lunar motion were coming to a grinding halt.[21]

Curtis Wilson considered Horrocks's theory the chief improvement in lunar theory during the seventeenth and early eighteenth centuries, thereby implicitly giving it more weight than Newton's theory of gravitation and its – currently unclarified – influence on lunar theory: 'The [Newton's] approach [to lunar theory] failed, and as a predictive model, Horrocks's own theory remained as good as any available down to the publication of Tobias Mayer's

---

[20] On Horrocks, see: John Wallis' dedicatory letter to William Brouncker in Horrocks (1673), also Chapman (1982) and Wilson and Taton (1989, Chap. 10). His lunar theory is included in Wallis' publication of Horrocks's collected work (Horrocks 1673) and discussed in Wilson (1987). His idea of the lunar orbit as a form-changing ellipse was seminal to Euler's later work concerning variation of constants.

[21] Whiteside (1976).

first lunar tables in 1753.'[22] Wilson's view is endorsed by George Smith: 'Newton himself did not significantly advance the problem of the moon's motion beyond Horrocks.'[23] On the other hand, Smith *does* venerate Newton's significant contribution of providing the study of the moon's motion with a gravitational basis. We must also remind that more than half of the longitude equAtions in *NTM* were original and at least qualitatively, if not quantitatively, derived from the law of gravitation. We should not, therefore, interpret *NTM* as only an improved version of Horrocks's lunar theory. One might say that Newton's was the first to contain equAtions *predicted* to be of any relevance, without first having been *observed*.[24]

6.4 Old Wine in New Bottles

We will now translate the kinematic Horrocksian model of the previous section into the language of trigonometric functions. It will become clear that quite a number of other equAtions are involved. Mayer, apparently having an interest in Newton's *NTM*, made computations comparable to those below. A brief comment on his calculations will be made near the end of this section. Later in the chapter, his calculations will also be seen to play a crucial role in the conception of the kil tables.

First, we need to analyse how the Horrocksian variable orbit affects the equAtion of centre. For an unperturbed elliptical motion, the equAtion of centre $\mathcal{C}$ may be written to the third order in the eccentricity e as

$$\mathcal{C} = -\left(2 - \tfrac{1}{4}e^2\right)e \sin v + \tfrac{5}{4}e^2 \sin 2v - \tfrac{13}{12}e^3 \sin 3v, \qquad (6.1)$$

where v denotes the mean anomaly measured from the apogee. The key concept of Horrocks's variable orbit is to periodically change the eccentricity and the apse line orientation of the lunar orbit, and to take the equAtion of centre pertaining to the form-changing instantaneous ellipse. Referring to either Fig. 6.1 or 6.2, we put the mean eccentricity $TC = a$, the actual eccentricity $TF = e$, and the fixed ratio $FC : TC = \beta$. The kinematic model specifies that $\angle FCB = 2\angle STC = 2(\omega - p)$, where p equals the lunar mean anomaly (once equAted for the annual equAtion of the apogee), and ω the mean lunar longitude minus mean solar longitude. Hence we express the eccentricity via

$$e^2 = TF^2 = TC^2 + FC^2 + 2TC \cdot FC \cos \angle FCB$$
$$= a^2 + a^2\beta^2 + 2a^2\beta \cos(2\omega - 2p). \qquad (6.2)$$

---

[22] Wilson (1987, p. 77); also see Wilson and Taton (1989, p. 194).

[23] Newton et al. (1999, p. 256).

[24] Wilson (1987, p. 77), Newton et al. (1999, p. 256); also see Wilson and Taton (1989, p. 194).

For the modified anomaly in (6.1) we write

$$v = p + \delta,$$

where, deviating from our usual meaning of the symbol δ for the duration of this section, $\delta = \angle FTC$, the second equAtion of the apogee. Let D (not in figure) be the foot of the perpendicular from F to TC. The following relations hold for δ:

$$\sin \delta = FD/TF, \quad \text{whence} \quad e \sin \delta = a\beta \sin(2\omega - 2p),$$
$$\cos \delta = TD/TF, \quad \text{whence} \quad e \cos \delta = a(1 + \beta \cos(2\omega - 2p)). \quad (6.3)$$

Equations (6.2) and (6.3), which provide the moon's variable eccentricity as well as the second equAtion of the apogee, will be of use again in the next section. Presently, we use them to write the first-order term in the modified equAtion of centre as

$$\begin{aligned} e \sin(p + \delta) &= e \cos \delta \sin p + e \sin \delta \cos p \\ &= a(1 + \beta \cos(2\omega - 2p)) \sin p + a\beta \sin(2\omega - 2p) \cos p \\ &= a \sin p + a\beta \sin(2\omega - p). \end{aligned} \quad (6.4)$$

This is beautiful. The term $a \sin p$ that shows up on the right-hand side is the first-order term of the unperturbed equAtion of centre for the mean eccentricity a and mean anomaly p. The second term has argument $2\omega - p$ and we recognize it as the prime evection term. Thus we learn the important fact that the Horrocksian-modified equAtion of centre equals the sum of the unperturbed equAtion of centre and the evection, to the first order in the eccentricity. Next, we combine (6.2), (6.4), and the familiar relation

$$\sin x \cos y = \tfrac{1}{2}(\sin(x + y) + \sin(x - y)), \quad (6.5)$$

in order to expand the first term of (6.1) into

$$\begin{aligned} (2 - \tfrac{1}{4}e^2) e \sin(p + \delta) &= (2a - \tfrac{1}{4}a^3 - \tfrac{1}{2}a^3\beta^2) \sin p \\ &\quad + (2a\beta - \tfrac{1}{2}a^3\beta - \tfrac{1}{4}a^3\beta^3) \sin(2\omega - p) \\ &\quad - \tfrac{1}{4}a^3\beta^2 \sin(4\omega - 3p) + \tfrac{1}{4}a^3\beta \sin(2\omega - 3p). \end{aligned} \quad (6.6)$$

Continuing now to the second term in (6.1), we first observe that

$$e^2 \sin 2\delta = 2e^2 \sin \delta \cos \delta$$
$$= 2a^2\beta \sin(2\omega - 2p)\big(1 + \beta \cos(2\omega - 2p)\big)$$
$$= 2a^2\beta \sin(2\omega - 2p) + a^2\beta^2 \sin(4\omega - 4p),$$
$$e^2 \cos 2\delta = e^2(\cos^2 \delta - \sin^2 \delta)$$
$$= a^2\big(1 + \beta \cos(2\omega - 2p)\big)^2 - a^2\beta^2 \sin^2(2\omega - 2p)$$
$$= a^2 + 2a^2\beta \cos(2\omega - 2p) + a^2\beta^2 \cos(4\omega - 4p);$$

hence

$$e^2 \sin(2p + 2\delta) = e^2 \sin 2p \cos 2\delta + e^2 \cos 2p \sin 2\delta$$
$$= \sin 2p\big(a^2 + 2a^2\beta \cos(2\omega - 2p) + a^2\beta^2 \cos(4\omega - 4p)\big)$$
$$\qquad + \cos 2p\big(2a^2\beta \sin(2\omega - 2p) + a^2\beta^2 \sin(4\omega - 4p)\big)$$
$$= a^2 \sin 2p + 2a^2\beta \sin 2\omega + a^2\beta^2 \sin(4\omega - 2p). \qquad (6.7)$$

We see here that the second-order term of the modified equAtion of centre contributes to the variation and also (in the second order) to the evection. Repeating the same procedure for the third term in (6.1) yields

$$e^3 \sin(3p + 3\delta) = a^3 \sin 3p + 3a^3\beta \sin(2\omega + p)$$
$$\qquad + 3a^3\beta^2 \sin(4\omega - p) + a^3\beta^3 \sin(6\omega - 3p). \qquad (6.8)$$

Finally then, we find that the equAtion of centre (6.1) expands into

$$C = \left(-2a + \tfrac{1}{4}a^3 + \tfrac{1}{2}a^3\beta^2\right)\sin p + \tfrac{5}{4}a^2 \sin 2p - \tfrac{13}{12}a^3 \sin 3p + \tfrac{5}{2}a^2\beta \sin 2\omega$$
$$+ \left(-2a\beta + \tfrac{1}{2}a^3\beta + \tfrac{1}{4}a^3\beta^3\right)\sin(2\omega - p)$$
$$+ \tfrac{5}{4}a^2\beta^2 \sin(4\omega - 2p) - \tfrac{13}{12}a^3\beta^3 \sin(6\omega - 3p)$$
$$- \tfrac{1}{4}a^3\beta \sin(2\omega - 3p) + \tfrac{1}{4}a^3\beta^2 \sin(4\omega - 3p)$$
$$- \tfrac{13}{4}a^3\beta^2 \sin(4\omega - p) - \tfrac{13}{4}a^3\beta \sin(2\omega + p).$$

$$(6.9)$$

Upon substitution of the typical Newtonian values $a = 0.05505$ and $\beta = 0.2131$, and rounding to arc-seconds, we find that the equAtion of centre in the form-changing ellipse equals

$$C = -\,6°18'20'' \sin p + 13'1'' \sin 2p - 37'' \sin 3p + 5'33'' \sin 2\omega$$
$$- 1°20'36'' \sin(2\omega - p) + 35'' \sin(4\omega - 2p)$$
$$- 2'' \sin(2\omega - 3p)$$
$$- 5'' \sin(4\omega - p) - 24'' \sin(2\omega + p).$$

$$(6.10)$$

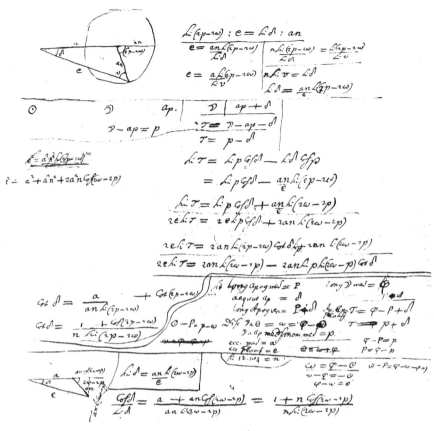

Fig. 6.3 Part of an unnumbered folio of Cod. μ^{H}_{28}, showing Mayer's work on the Horrocksian mechanism. The top half of the folio contains an error, which is corrected in the bottom half (below the double line)

Several authors have presented a similar but less extensive analysis.[25] In particular, Gaythorpe and Wilson showed that the Horrocksian variable orbit is equivalent to the combined equAtion of centre and evection. But Gaythorpe missed the contribution to variation of $\approx 5' \sin 2\omega$ in (6.9). Because of this contribution, every Horrocksian lunar theory (i.e., with variable eccentricity and apsidal line plugged into the equAtion of centre) necessarily has a variation coefficient of $\approx 35'$ instead of the total $\approx 40'$.[26]

Mayer too made computations such as those above, as I discovered on various loose folio sheets among his work on lunar theory (cf. Fig. 6.3). The same sheets carry several other computations that are manifestly connected

---

[25] These include d'Alembert (1754–1756, I pp. 91–93), Godfray (1852, p. 60), Gaythorpe (1957), Brown (1896), and Wilson and Taton (1989, p. 198).

[26] Gaythorpe's oversight was corrected by Jørgensen (1974); the point had, however, already been noted by d'Alembert (1754–1756, I p. 93, 253).

with his endeavours to understand the successive steps of Newton's lunar theory in the language of trigonometric quantities. In these computations, Mayer kept terms involving $a^3\beta^2 \approx 2''$, and rejected $a^3\beta^3 \approx 0.3''$ and a^4. Because $\beta^2 \approx a$, he might have rejected $a^3\beta^2$ just as well. He treated the trigonometric quantities in a distinctly algebraical style, as developed by Euler since 1739,[27] and very similar to our treatment on the preceding pages of this section.

The contrast between Newton's geometrical style and Mayer's reformulation in trigonometric quantities is telling of the change in perception of trigonometry which Euler had brought about. This contrast is perhaps even more clearly perceived when we turn to the lunar tables: first to those of Lemonnier, which are fully based on Newton's *NTM*, then to the twist that Mayer gave to Lemonnier's tables.

6.5 Lemonnier's Version of *NTM*

Although Flamsteed did not publish his *NTM*-based lunar tables, some form of a copy of his manuscript must have reached France, where Lemonnier adapted it for publication in his *Institutions Astronomiques*,[28] an enlarged translation of John Keill's *Introductiones ad veram physicam et veram astronomiam*. Mayer was evidently well acquainted with both books.[29]

Because *Institutions Astronomiques* was an important source to Mayer, we will now undertake to appraise the conformity of Lemonnier's tables with Newton's precepts. Our general approach is as follows. Because Lemonnier

---

[27] Euler expounded the calculus of the trigonometric functions in his treatise on the great inequality of Jupiter and Saturn (Euler 1749a), which Mayer knew very well. See Katz (1987, p. 322), Golland and Golland (1993).

[28] Lemonnier and Keill (1746), also see comments in Kollerstrom (2000, pp. 205–214).

[29] Forbes (1971a, p. 83), Forbes (1980, p. 120). Rob van Gent suggested (private communication) that an interesting alternative source of information of Flamsteed's tables to Mayer might have been Johann Gabriel Doppelmayr (1671–1750). Both Germans were working together in the Homann building during the later 1740s. Doppelmayr visited England in 1701, where he met Gregory for certain, and Flamsteed and Newton very probably (Gaab 2001). Newton had then already written his *NTM* (Kollerstrom 2000, p. 43). Gregory had not yet published it but at least he might have known it. In 1705, after returning to Nuremberg, Doppelmayr published his Latin translation of Streete's *Astronomia Carolina*. Although Streete's work was surely a long-standing classic in astronomy, one might expect Doppelmayr to include the Flamsteed lunar tables instead of Streete's – which he did not, suggesting that the newer Newtonian tables made at most no lasting impression on him. Moreover, Flamsteed calculated his manuscript lunar tables of the Newton variety only after Doppelmayr had left England, and he apparently made only two manuscript copies of them (Baily 1835, p. 695, 704). The text on the lunar plate of Doppelmayr's *Atlas Coelestis* (1742) suggests that the author did not fully grasp the details of *NTM*. Strikingly, though, Mayer consistently referred to the lunar tables of *Flamsteed*, never to those of *Lemonnier*, in his manuscripts, perhaps because Lemonnier had dutifully acknowledged Flamsteed as the provenance of his tables.

Display 6.1 EquAtions of lunar longitude, node, and inclination of Lemonnier's lunar tables

1	Annual equAtions	
	to longitude	$+11'49'' \sin \varsigma - 7\frac{1}{2}'' \sin 2\varsigma$
	to apogee	$-20'\ 0'' \sin \varsigma + 13'' \sin 2\varsigma$
	to node	$+\ 9'30'' \sin \varsigma -\ 6'' \sin 2\varsigma$
2		$(+3'45'' - 11'' \cos \varsigma) \sin(2\omega - 2p)$
3		$+47'' \sin(2\omega - 2\delta)$
4	2nd apogee equAtion	$\arctan \dfrac{\beta \sin(2\omega - 2p)}{1 + \beta \cos(2\omega - 2p)}$
	eccentricity e	$a\sqrt{1 + \beta^2 + 2\beta \cos(2\omega - 2p)}$
	equAtion of centre	$(2e - \frac{1}{4}e^3) \sin p - (\frac{5}{4}e^2 - \frac{11}{24}e^4) \sin 2p$
		$+\frac{13}{12}e^3 \sin 3p - \frac{103}{96}e^4 \sin 4p$
5	Variation	$(35'15'' - 2'11'' \cos \varsigma) \sin 2\omega$
6		$+2'10'' \sin(2\omega - p + \varsigma)$
7		$-2'20'' \sin \omega$
	2nd node equAtion	$\arctan \dfrac{\sin(2\delta - 2\omega)}{\gamma + \cos(2\delta - 2\omega)}$
	Eqn. of Inclination	$5°8'30'' + 9' \cos(2\delta - 2\omega)$

The leftmost numbers correspond to the steps in Newton's *NTM*, described in section 6.2

acknowledged Flamsteed as his source, we may assume as a working hypothesis that the form of the equAtions on which his tables are based agrees to the Newtonian prescriptions as discussed above. We then deduce the coefficients of these equAtions from the values in Lemonnier's tables. This is straightforward for some equAtions, but more elaborate for others. Finally, we check if the equAtions and coefficients indeed reproduce Lemonnier's tables sufficiently accurate, i.e., to within $1''$ or $2''$.

We now turn to the details, referring to Sect. 6.2 for Newton's seven steps, and to Display 6.1 for the results of the analysis, and, as usual, using the argument notation of Mayer.

Newton's first step comprises three annual equAtions, proportional to the solar equAtion of centre. The latter is a formula of the form $c_1 \sin \varsigma + c_2 \sin 2\varsigma + c_3 \sin 3\varsigma + \cdots$, where ς is the solar mean anomaly, and the coefficients stand in the ratio $c_1 : c_2 : c_3 = 95 : 1 : \frac{1}{69}$, roughly. The three annual equAtions should then take the same form, with the same ratio between the coefficients, while the coefficients follow from the extreme values of the equAtions as specified by Newton. This yields the annual equAtions listed in Display 6.1.[30] Lemonnier's tables satisfy these equAtions.

---

[30] The third coefficients of each are negligible, cf. fn. 13 on p. 95.

Lemonnier's second equAtion is equivalent to Newton's second step, and may be represented as $(3'45'' - 11'' \cos \varsigma) \sin(2\omega - 2p)$. As explained in Sect. 6.2, this equAtion has a seasonal fluctuation in its coefficient. Lemonnier has two tables to represent this equAtion: one table gives $3'45'' - 11'' \cos \varsigma$ as a function of ς, the other tabulates $3'45'' \sin(2\omega - 2p)$. The user of these tables has to perform an additional calculation to establish the magnitude of the second equAtion. If, for given values of the arguments ς and $2\omega - 2p$, the first table yields x and the second table returns y, then $xy/3'45''$ is the value of the second equAtion.

A similar seasonal fluctuation is also found in Newton's step 5, his variation, which we have represented previously as $(35'32'' - 1'53'' \cos \varsigma) \sin(2\omega)$. Lemonnier's fifth equAtion takes two tables again, but his coefficients differ slightly from Newton's values: the table for the seasonal part matches $35'15'' - 2'11'' \cos \varsigma$, the other table lists $35'15'' \sin(2\omega)$.[31]

The third and sixth steps of Newton are easy to implement in tables. For these equAtions, a straightforward check confirms that Lemonnier apparently adhered to Newton's coefficients of *NTM*. He also tabulated Newton's seventh equAtion except that he dropped its seasonal term.

Finally, we come to Newton's fourth step, with its variable eccentricity and oscillating apsidal line in the Horrocksian way. Lemonnier implements the motion of the apsidal line as a second equAtion of the apogee, but the eccentricity has a more involved rendering. We will illuminate the apogee equAtion first.

In Fig. 6.1 and equation (6.3), $\angle FTC = \delta$ is the second apogee equAtion, and $FC : TC = \beta$. It follows from (6.3) that

$$\delta = \arctan \frac{\beta \sin(2\omega - 2p)}{1 + \beta \cos(2\omega - 2p)}. \tag{6.11}$$

Presuming that Lemonnier tabulated this relation, with $\omega - p$ as the argument, we proceed as follows to estimate β from his table. Solving the last equation for β yields

$$\frac{1}{\beta} = \frac{\sin(2\omega - 2p)}{\tan \delta} - \cos(2\omega - 2p). \tag{6.12}$$

Now we fill in all tabulated argument–value pairs $(\omega - p, \delta)$ to obtain just as many estimates for β, which we average to find our final estimate $\bar{\beta} = 0.213104$. This $\bar{\beta}$ substituted back in (6.11) suffices to recreate Lemonnier's table with no differences larger than $2''$. Besides, this estimate agrees with the ratio for $FC : TC = \beta$ that we have computed on p. 99 from data in *NTM*. Thus, we may conclude that Lemonnier's second apogee equAtion is in perfect agreement with *NTM*.

---

[31] Lemonnier combined the seasonal tables of equAtions 2 and 5 in one table, for practical reasons.

To account for the variable eccentricity, Lemonnier again included two tables. The first of these listed the extremal values of the equAtion of centre for different values of the argument $\omega - p = \angle STC = \frac{1}{2}\angle FCB$ in Fig. 6.1. The second table listed four different equAtions of centre, with, respectively, extremal values of $-5°$, $-6°$, $-7°$, and $-7°39.5'$. The user was then supposed to interpolate (or even extrapolate) between two columns of this table, depending on the extremal value produced by the first table. Confusingly, the headings of the second table contained not these extremal values, but the eccentricities at which they occur. The link to the extremes was only explained in a commentary.[32]

Lemonnier's table for maximum equAtion of centre can be reproduced to within $3''$, as follows: compute the variable eccentricity e according to (6.2), with a mean eccentricity of $a = 0.05506$ instead of Newton's value 0.05505; then substitute this eccentricity in the equAtion of centre, and compute the equAtion for argument $p \approx 94°$.[33]

With the well-known formula for the equAtion of centre, Lemonnier's fourfold table of that function is then reproducible to $2''$ or better. Interestingly, Lemonnier had apparently included terms of the fourth order in the eccentricity (see formula in the middle of Display 6.1).

This completes our analysis of Lemonnier's lunar tables. Display 6.1 also shows their equAtions for node and inclination; these will not be further discussed here.[34] Kollerstrom rightly remarked that Lemonnier's tables form a true representation of *NTM*, with a few small changes to parameters, without the seasonal modulation of the 7th equAtion, and with the sign error of *NTM*'s sixth equAtion corrected.[35] D'Alembert published a similar analysis, but less detailed, missing for example the second terms in the annual equAtions.[36]

On the practical side, we have seen that Lemonnier's tables necessitate multiplications and divisions to accommodate the seasonally varying equAtions, and that they incorporate a somewhat complicated scheme to accommodate the variable eccentricity. A user of Lemonnier's tables has to make more involved calculations than a user of Mayer's tables, as exemplified in Chap. 4.

---

[32] Lemonnier and Keill (1746, p. 629). The link between argument $\omega - p$ and eccentricity e is provided by (6.2) above.

[33] The equAtion of centre attains its maximum when $\cos p = (5e)^{-1} - \sqrt{(5e)^{-2} + \frac{1}{2}}$; for the eccentricities of interest this maximum is reached when $p \approx 94° \pm 1°$.

[34] The second node equAtion takes $\gamma = 38.3341$. These equAtions were demonstrated by Newton in the *Principia*. They do not affect lunar longitude.

[35] Kollerstrom (2000, p. 212).

[36] d'Alembert (1754–1756, I, Chap. 13).

6.6 Lunar Tables, Probably Older?

Among the Mayer manuscripts in Göttingen is the quire Cod. $\mu_{15}^{\sharp}$, to which Lichtenberg added the title '*Mondtafeln (wahrscheinlich älterer Entwurf)*' ('Lunar tables, probably of older design'). It is of interest here because it is a witness of the introduction of multiple steps in Mayer's lunar tables. Most of the quire was composed during 1752. It marked the transition from Mayer's single-stepped zand theory, contained in a letter to Euler of 1752 January 6, to the multistepped precursors of the kil tables published in the spring of 1753.[37] Part of its contents depended on a design considerably older than Lichtenberg might have perceived, as we will see.

The items in the manuscript that are currently of interest are the following: (1) a comparison of the coefficients in several lunar theories on pp. 8v and 9r; (2) tables of lunar equAtions on pp. 9v–16v; (3) a page (p. 17r) with the superscript *Entwurf neuer ☽ Tafeln*; followed by (4) again tables of lunar equAtions. Each of these are discussed below, whereafter I provide an interpretation of the manuscript.[38]

6.6.1 Peering at the Peers

The facing folios 8v and 9r of Cod. $\mu_{15}^{\sharp}$ are laid out in the form of an array (see Fig. 6.4). The first column on the left side of the array lists trigonometric expressions: successively $\sin p$, $\sin 2p$, $\sin 3p$, $\sin \varsigma$, ..., $\sin(p - \varsigma)$, $\sin(p + \varsigma)$, ... The other columns bear the following superscripts (numbers added in square brackets for ease of reference): [1] *Clairaut*, [2] *Calculus m. ex theor. M.*, [3] *Corr. ex observ. M.*, [4] *Eul.*, [5] *Tabb ☽. Calc. M.*, and [6] *New. Flamst.* Under these headings, the columns are filled with numbers, which are clearly coefficients, forming equAtions together with the trigonometric quantities on the left side.

The coefficients in column [3] agree exactly to zand, i.e., to those that Mayer transmitted to Euler in his letter as mentioned above. In this letter, which provides an excellent backdrop against which Cod. $\mu_{15}^{\sharp}$ falls into perspective, Mayer explained why he chose to base his tables on the *mean* arguments, contrary to Euler's preference for the *eccentric* arguments:

> The angles ω, p, and ς invariably denote the mean motion, which in fact brings several advantages not in the solution, but in practice, because in such a manner the arguments of the inequAlities can be calculated more simply.[39]

Next in the same letter he referred to the problem of the motion of the lunar apogee. Although the famous problem of its *mean* motion had recently been

---

[37] Letter to Euler: Forbes (1971a) p. 48.

[38] Aliases of the versions treated here are all listed in Display A.1.

[39] Forbes (1971a, p. 49).

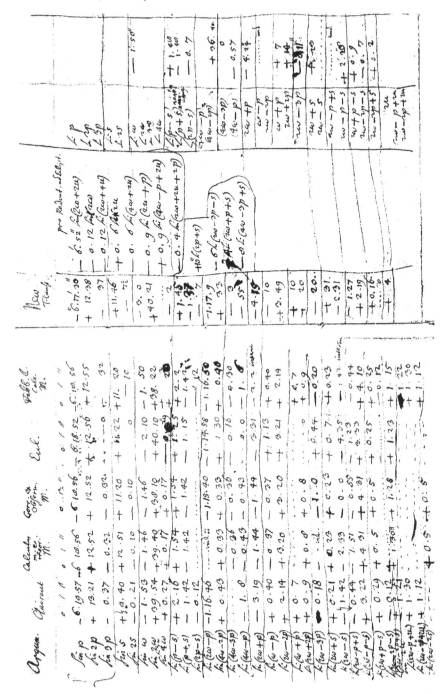

Fig. 6.4 Comparison of coefficients from various astronomers, on fol. 8v–9r in Cod. $\mu_{15}^{\sharp}$

solved by Clairaut, Mayer's remark makes more sense when interpreted in relation to the *variable* part of the motion of the apogee, $\angle FTC$ in Fig. 6.1:

> I have indeed always supposed, yet could never be certain, that the inequality associated with the angle $2\omega - p$ [i.e., evection], which is without doubt difficult to determine and which was explained by Newton as due to the variation of the eccentricity of the Moon's orbit, strongly affects the motion of the apogee. Now, I want particularly to make new attempts to determine the motion of the apogee, and to see whether I do not arrive at it if instead of the above-mentioned inequality I take the eccentricity as being really variable.[40]

In fact, this quotation is intelligible only in the context of Newton's lunar theory and the Horrocksian variable orbit, even though it presages the multistep tables. In the same letter, Mayer had indicated his desire to see Euler's and Clairaut's lunar theories:

> Meanwhile, I eagerly await the treatises of yourself and Mr. Clairaut. You would greatly oblige me if I could obtain through your assistance a copy as soon as they are published....[41]

Euler had been an adjudicator for the 1751 prize contest of the Academy of Saint Petersburg on lunar theory, which was won by Clairaut's contribution, therefore Euler was an eminent source of information. He responded on 18 March 1752 that he was unable to compare the magnitudes of the equAtions of his own lunar theory with Mayer's, because of the differences in the arguments already alluded to: Euler's tables made use of the eccentric anomaly, but Mayer's used the mean anomaly throughout. Like Mayer's, Clairaut's equAtions used the mean arguments, and therefore Euler included a complete list of the equAtions in Clairaut's theory. Euler kept the equAtions in the same sequence as Mayer had in his letter, inserting a few that were present in Clairaut's, but absent from Mayer's theory (most notably several equAtions that involved the angular distance of the sun from the nodes of the lunar orbit, which Mayer overlooked until then). Mayer copied precisely this list out of Euler's letter into column [1] of the array on fol. 8v and 9r of Cod. $\mu_{15}^{\sharp}$. Therefore, we can be sure that Mayer started this array after he received Euler's response; moreover, we may speculate that he did not waste much time before doing so.

Mayer's own coefficients in columns [2] and [3] are mostly identical: apparently, he adjusted only a few of his theoretically derived coefficients in the former to observations. The superscript of column [4] suggests that Mayer included coefficients from Euler's tables. I have not investigated how Mayer obtained these; because of the difference in arguments, it was probably a non-trivial exercise.

Column [5] is peculiar in that the coefficients in it sometimes follow Mayer's values, sometimes those of Clairaut (particularly for those equAtions where

---

[40] Forbes (1971a, p. 49).

[41] Forbes (1971a, p. 49).

Mayer did not have a value of his own), and sometimes they fall in between. Only one coefficient was drawn from Euler's column, and two are marked as preliminary ('*interim*'). It seems as if Mayer was, in a sense, interpolating between his own, Clairaut's, and (to a lesser extent) Euler's lunar theories.

Column [6] with its superscript '*New. Flamst.*' suggests that it is linked to Newton's lunar theory. Indeed it is: Mayer calculated these coefficients on the same folios of Cod. $\mu_{28}^{\sharp}$ where he expressed Horrocks's variable eccentricity and apsidal line as trigonometric quantities, as mentioned on p. 103 above. Judging by the layout, this column was added perhaps a little while after the other columns had been completed.

6.6.2 The First Set of Lunar Tables

The next item of interest in Cod. $\mu_{15}^{\sharp}$ is a set of lunar tables aliased grond. It consists of mean motion tables equivalent to Lemonnier's, (except that their node epoch had been adjusted), followed by 19 tables of equAtions catering for exactly the equAtions as listed in column [5], fol. 8v, with the same coefficients. The superscript *Tabb ☽. Calc. M.* of column [5], just discussed, must refer to these tables.

6.6.3 Design of New Tables

Apparently Mayer abandoned those grond lunar tables; they are immediately followed by a page so revealing that I have transcribed it in Display 6.2. The following observations apply.

A heading *Entwurf neuer ☽ Tafeln* is written along the top edge of the page: 'design of new lunar tables'. The design is specified below the heading. There are 11 numbered arguments, which are given symbolically as well as in descriptive language. For each argument, an equAtion is specified, with one or more terms, and appropriate coefficients in sexagesimal notation. The majority of the equAtions apply to lunar longitude, but there are a few that apply to the apogee or node.

The plain-language descriptions of the arguments in the third column clearly reveal that some of them are to be corrected in steps. For example, the elongation of the moon is to be corrected using the first equAtions, depending on argument ς, the mean solar anomaly. Mayer indicates this by his addition '*I corr.*' in the description of arguments II and V. Similarly, argument IV is the anomaly of a 'corrected' moon; Mayer uses the symbol q to distinguish it from the mean anomaly p. There is a similar distinction between the elongations ω and $\tilde{\omega}$, although less consistently applied: argument VI is lunar elongation after application of the V equAtion (and, I suppose, equAtions I to IV as well).

To summarize, Mayer introduces the multistep procedure here in this 'new
design'. By the doubled lines between III and IV and between V and VI, he
clearly distinguishes the three steps.

The basic structure of this design has striking resemblances to Newton's
NTM. To wit, the first argument (solar mean anomaly ς) is not only used to
equAte the lunar *longitude*, but also its apogee and node positions, exactly
as in the 1702 theory. At least as telling, we discern the hallmark of *NTM*
in Mayer's design: the equAtion of centre and the evection close together
as equAtions IV and V in the middle step. Both recipes have the variation
as their next equAtion; Mayer's subsequent equAtions take up its seasonal
modification prescribed by Newton. The resemblance of the two designs is
certainly more than coincidental. Differing from *NTM*, Mayer's design collects
several equAtions together into one step. Mayer also has appended two new
equAtions, but apparently he dropped them almost immediately afterwards.
Most (but not all, typically) of his coefficients differ from Newton's.

6.6.4 The Second Set of Lunar Tables

On the following pages of the manuscript we find lunar tables in an untidy
handwriting, as if they were a preliminary or intermediate version. Apart
from mean motion tables and an unnumbered equAtion of centre answering
to $-6°18'56'' \sin p + 12'55'' \sin 2p - 32'' \sin 3p$, we find:

II	$+40'' \sin(\omega - p) + 3'20'' \sin(2\omega - 2p)$
III	$-58'' \sin(2\omega + \varsigma)$
IV	$-58'' \sin(2\omega - \varsigma)$
V	$-58'' \sin(2\omega - 2\delta)$
VI	$+1'30'' \sin(2\omega - p + \varsigma)$
VII	$+1'30'' \sin(2\omega - p - \varsigma)$
VIII	$+58'' \sin(2\delta - p)$
IX	$+30'' \sin(2\delta - 2p)$
X	$+1'30'' \sin(2\omega + p)$

At first sight, this does not seem to tally with the multistepped development
of the *Entwurf*. However, using techniques described in Sect. 8.3, I analysed
certain position calculations in Cod. $\mu_{41}^{\sharp}$, and discovered that they used the
equAtions embodied in these tables, and moreover that these calculations
adhered to a computational scheme with multiple steps.[42] Unlike what the

---

[42] Those calculations on fol. 1–29, 74, 90, 91 apparently used equAtions much like the ones
at hand, and from the context it appears that Mayer is working to improve his tables. His
work included improvement of the mean motions, from those of the first table set mentioned
on p. 111 above, to those of the published kil tables. Mayer compared the outcome of his
table computations with observations of James Bradley that Euler had sent him in the fall
or winter of 1752 (Forbes 1971a).

Display 6.2 *Design of new lunar tables*, transcription of fol. 17r in Cod. $\mu_{15}^{\#}$, alias *geer*. The small numbers 1 and 2 in the *left margin* and the dots after q in Arg. IV are as in Mayer's original. The illegible coefficient in eqn. VII should read 2′10″

Entwurf neuer ☽ tafeln.

			Aeq. ☽	Aeq. apog.	Aeq. Ω
1 Arg.I.	ς	Anomalia media Solis	$+11'40''\sin\varsigma$ $-\;10''\sin 2\varsigma$	$-20'0''\sin\varsigma$ $+\;20''\sin 2\varsigma$	&c
Arg.II.	$\omega-p$	Dist ☽ a ⊙ I corr. − Anom ☽ med	$+3°20''\sin(2\omega-2p)$ $\pm\,20\sin(\omega-p)$		
2 Arg.III.	u	Long ⊙ − Long Ω med.	$-1'0''\sin 2u$		
Arg.IV.	$q\dots$	Long ☽ corr. −Ap.☽ corr.	$-6°18'27''\sin q$ $+\;12.38\sin 2q$ $-\qquad 37\sin 3q$		
Arg.V.	$2\omega-q$	ad arg II add. Dist.☽ a ⊙ I.corr.	$-1°21'0''\sin(2\omega-p)$ $+\quad 36\sin(4\omega-2p)$		
Arg.VI.	$\tilde{\omega}$	Dist. ☽ a ⊙ post V aequationem	$-2'0''\sin\omega$ $+40.21\sin 2\omega$ $+\quad 2\sin 3\omega$ $+\quad 22\sin 4\omega$		
Arg.VII.	$2\tilde{\omega}-q+\varsigma$	ad arg V add. anom ⊙ med	$+[?].10\sin(2\omega-q+\varsigma)$		
Arg.VIII.	$2\omega+\varsigma$	ad arg VII add anom☽ add arg IV	$-1.10\sin(2\omega+\varsigma)$		
Arg.IX.	$2\omega-\varsigma$	ad dupl dist ☽ a ⊙ subt an ⊙	$-0.30\sin(2\omega-\varsigma)$ vel 1.0 vel 0.0		
Arg.X.	$2\omega-q+2u$	ad Arg V add dupl.arg III.	$+1.0\sin(2\omega-q+2u)$		
Arg.XI.	$2\omega-2q+2u$	ab Arg X subtr. Arg IV	$+0.30\sin(2\omega-2q+2u)$		

numbering suggests, their first equAtion is an annual equAtion; the second step is equAtion of centre and evection, and the third step is variation. Thus, the computations show that these tables play a role in a procedure identical to the zwin version, which Mayer wrote to Euler on the 7th of January, 1753, and to the kil tables, which were published in the spring of that year.[43]

6.6.5 Interpretation

In short, the manuscript Cod. $\mu_{15}^{\sharp}$ contains a list of the coefficients of Clairaut's lunar theory, compared with coefficients of Mayer's own theory at that time (1752), as well as to their fitted counterparts of gors-zand, and also to coefficients extracted from theories of Euler and ultimately (perhaps added a while later) of Newton-Flamsteed-Lemonnier. This is followed first by tables based on a kind of mediated coefficients, then by a sketch of the new design that adopted Newton's multistepped scheme, and finally by tables of a multistepped nature, which can be regarded as predecessors of the kil tables.

This and the letters exchanged between Euler and Mayer lead us to the following interpretation. Prior to 1752, Mayer had a lunar theory, and he had fitted its coefficients to observations. Not being satisfied with the result, he discussed various aspects of lunar theory with Euler, including a variable eccentricity and the uneven motion of the apsides. Euler kindly transmitted the coefficients of Clairaut's theory. Mayer looked closely at the results of his fellow mathematicians and mingled some of them with his own coefficients, but soon he gave up and turned to a Newtonian-like scheme characterized by multiple steps, annual equAtions for node and apogee, and an analytical equivalent of Horrocks's variable ellipse. This marked the start of a new branch of development. Strikingly, Mayer reverted to Newton's theory like Newton had reverted to Horrocks's.

6.7 Accuracy of Theories Compared

To assess if Mayer reached an improved accuracy by adopting the Newtonian steps, I conducted the following numerical experiments. I loaded several sets of (Mayer's and Lemonnier's) coefficients into a computer program. Each set corresponded to a different version of tables; and with each set the program computed 1,000 lunar longitudes at 3-day intervals starting 1 January 1740,

---

[43] In fact, it appears that the development went from the geer *Entwurf* via gat and put to zwin, which is almost identical to its successor kil. Improvements of put and kil are discussed in Chap. 8.

covering a little less than half a Saros. These longitudes were subtracted from ones obtained by a modern theory, and the standard deviation of the differences was taken. The results are listed in Display 6.3. All the non-modern lunar longitudes were computed using a modern solar theory,[44] and all computations were conducted with one and the same version of Mayer's lunar mean motion parameters. Therefore, the computations usually do not reproduce the same lunar longitudes as Mayer would have obtained himself. This is perfectly reasonable because our goal here is to get an impression of the error distribution of positions generated from various of Mayer's periodic equAtions. As it turns out, the adopted mean-motion practice has an effect primarily on the mean of the computed differences, and hardly on their standard deviations. By using the same solar theory and lunar mean motions with various equAtion versions, we concentrate on the quality of those equAtions.

Kollerstrom reported that Newton's lunar theory, with the sign error in the sixth equAtion corrected, had a standard deviation of $\sigma = 1.88' \approx 113''$ on 40 samples at 4-day intervals after 1681.0 using his full computer implementation of *NTM*. I found $\sigma = 123''$, using a somewhat different implementation that had the variable eccentricity and apsidal line represented by trigonometric quantities obtained as in Sect. 6.4 (besides, the data ranges on which the statistics are based differ). I could discern that small variations to the mean motion parameters had an impact on the pattern of the position errors, but not on the standard deviation of the errors. Kollerstrom reported $\sigma = 1.9' = 114''$ for Lemonnier's tables, which agrees very well with my determination of $111''$. These correspondences provide some sense of feasibility of my approach. In various other sources, the accuracy of *NTM* has been quoted as anything ranging from $2'-3'$ (by Halley) to $8'-10'$ (by Flamsteed) but usually it was not exactly specified to what these numbers referred.[45]

Clairaut, in his lunar theory, included a list of 100 observations of the moon compared with positions predicted by his own tables, which also had been slightly adjusted to observations.[46] Interestingly, their standard deviation comes out at $110''$, suggesting that Clairaut's tables were no better than Lemonnier's. However, this conclusion must be treated with considerable care, because the data sets on which it is based are completely incomparable. A test of Clairaut's tables that was included in the *Connoissance des Temps* for the year 1783, suggests significantly smaller standard deviations.[47]

Display 6.3 shows that Mayer's **grond** version, being presumably an attempt to improve **zand** by mixing in his peer's results, did not meet its objective. But Mayer attained a dramatic fourfold increase in performance as soon

---

[44] Meeus (1998, pp. 163–165).

[45] Baily (1835, p. 695). Kollerstrom's numerical results are taken from Kollerstrom (2000, pp. 143–144, 227).

[46] Clairaut (1752b, p. 91).

[47] Lémery (1780); part of the difference can be explained by the circumstance that Lémery performed the test with the − fitted − tables of Clairaut's later (1765) edition of his theory.

Display 6.3 Standard deviations of the longitude terms in some table versions, computed for $n = 1,000$, starting date 1740 Jan. 1, step size 3 days, Xephem computer program taken as modern reference, computations done with the same (zwin) lunar mean motions and with modern solar mean motions; geer is the *Entwurf* version

alias	zand	grond	geer	gat	zwin	kil	rede	Lemonnier
$\sigma\;['']$	395	404	96	62	45	43	30	111

as he adapted the Newton–Lemonnier theory. We can imagine that Mayer felt he had discovered something to stick to, even when he lacked our modern statistical concepts.

When his kil tables went to the press less than a year later, he had again doubled the accuracy. In the decade between the 1752 zand variant and the final rede tables, Mayer gained a factor 13 in the accuracy of his equAtions, measured by their standard deviations. His changes in mean motion parameters have not been taken into account here.

6.8 The Preface to the kil Tables of 1753

Because I have just conjectured that the kil tables published in the Göttingen *Commentarii*[48] developed out of Mayer's embracing of *NTM*, it will be interesting to hold Mayer's own comments to those tables up against that conjecture, in particular with regard to their kinematic vs. dynamical background.

In his introduction to those tables, Mayer asserted that he had deduced the inequAlities out of 'that most famous theory of the great Newton', which Euler had first reduced to 'analytical equations', and which Mayer had himself – after several fruitless attempts – solved 'by a singular and sufficiently elegant method', although it would be too lengthy for him to explain how. Instead, he elaborated on the causes of the equAtions.[49]

---

[48] The tables were published, with an introduction, as Mayer (1753b). In the *Gentleman's Magazine* for August 1754 appeared an almost literal translation into English of their preface, which Forbes in turn included in Forbes (1980, pp. 143–146).

[49] 'Yet I have deduced these tables, so far as the inequAlities of motion are concerned, from that most famous theory of the great Newton; which the celebrated Mr. Euler has firstly reduced most elegantly to general analytical equations. In solving these equations, after trying other ways in vain, I have used a particular and quite elegant method, but to exhibit it here would take too long. Therefore I have resolved to disclose only those things that make it possible to see through the origin and causes of the inequAlities presented in the tables, so far as one can do so without calculation' (*'Deduxi autem has tabulas, quoad inaequalitates motuum, ex famosissima illa magni Newtoni theoria; quam Vir Celeberrimus* EULERUS *primus ad aequationes analyticas generales elegantissime reduxit. Usus sum in hisce aequationibus resolvendis post frustra tentatas alias vias methodo singulari satisque concinna, sed quam hic exponere nimis longum foret. Quapropter ea tantum indicare decrevi, quae*

From the vantage point of our current conjecture, we begin to wonder *which* of Newton's 'theories' he addressed: the theory of gravitation, or *NTM*? At first sight, his reference to Euler is with regard to the casting of Newtonian physics in the language of differential equations. But could not Mayer have had Euler's codification of trigonometric functions in mind, as a prerequisite to translate *NTM* into that formalism? Could his 'singular and elegant method' refer to the translation of the Horrocksian mechanism into the language of trigonometric functions? If so, that could explain why Mayer was silent about his method, for Newton's *NTM* might have been regarded as old-fashioned compared with the analytical advancement of Euler and Clairaut – even though its predictive accuracy was still first-rate. But Mayer's remark of finally attaining a solution after fruitless attempts seems to refer rather to gravitation and differential equations again. His message to the reader is of having deduced the tables from the Newtonian theory of gravitation. Yet, the *NTM*-like structure of the tables and the inadequate state of his theory at that time suggest otherwise.

Instead of supplying the reader with an account or even a sketch of his theory, Mayer chose rather to construe the nature of the various equAtions. After he had expounded both the equAtion of centre and the evection, he related these two equAtions to the variable eccentricity and apse movement as explained by Newton:

> By those XIth and XIIth equAtions [i.e., the equAtion of centre and evection], the same inequAlities are saved that Newton and those who have closely followed him have explained by a variable eccentricity of the lunar orbit and an unequal motion of the apsides. And although that method [of Newton] answers exactly to the theory, as indeed I can demonstrate, yet it is somewhat more difficult and almost useless for table calculation. The astronomy of the moon owes therefore much to the celebrated Euler. He has been the first who elegantly put a constant equAtion of centre linked with what we have called the evection, in the place of a variable eccentricity, and in this way he has furnished the theory of the moon, otherwise extremely intricate, with distinguished profit.[50]

Thus, Mayer showed in this text that he was well acquainted with *NTM*. He said that its variable orbit was not practical for lunar tables (as indeed our

---

ad originem causasque inaequalitatum in tabulis exhibitarum, quantum quidem sine calculo licet, perspiciendas facere possunt') Mayer (1753b). D'Alembert criticized that it was not Euler, but he himself and Clairaut, who had first produced analytical lunar theories (d'Alembert 1754–1756, I p. 252).

[50] 'Duabus istis aequationibus XI nempe & XII eaedem inaequalitates salvantur, quas Newtonus, & qui stricte eum secuti sunt, per variabilem eccentricitatem orbitae lunae & inaequalem apsidum motum explicuerunt. Et quamquam ista methodus theoriae, quod equidem demontrare possum, exacte respondet, paulo tamen est difficilior & ad calculum tabularem fere inepta. Multum igitur debet astronomia lunaris Cel[ebri] EVLERO. Is enim primus fuit, qui constantem aequationem centri iunctam cum ea quam evectionem diximus, loco eccentricitatis variabilis concinne substituit, eoque theoriam lunae alias maxime intricatam insigni compendio ornavit' (Mayer 1753b, p. 385); also see Forbes (1980, p. 143).

study of Lemonnier's tables confirmed) and he praised Euler for substituting a fixed orbit with a constant equation of centre: indeed, trigonometric functions enable the resolution of Horrocks's variable orbit into two fixed equations. But not a word about differential equations or forces! Strictly, the passage does *not* imply that there was no coherent theory supporting Mayer's tables. One wonders, though, in what way he saw Newton's variable eccentricity and apse as justified exactly *by theory* – obviously in the modern sense of the word, because in the old sense the Horrocksian motion forms *part of the theory*.

Continuing his tale, Mayer recognized that the variable distances between earth and sun, and also between moon and earth, give rise to periodic alterations of the evection and variation. For the evection, these alterations are expressed by equations depending on arguments $2\omega \pm \varsigma$, $2\omega - 2p$, and a part of the variation (argument 2ω). For the variation, they bring about equations depending on $2\omega \pm \varsigma$ and $2\omega \pm p$; the $2\omega - p$ effect being part of the evection. We have seen that some of these equations are implicitly present in *NTM*, in the form of coefficients varying annually with the mean anomaly of the sun. Mayer had no trouble to translate these into the form of his own tables.

In conclusion, Mayer offered very little in this preface to support his alleged dynamical theory. He carefully avoided to assert that he indeed had such a theory: his comments about it are evasive. On the other hand, the kinematics of *NTM* clearly show through.

6.9 Conclusion

I have shown that Mayer had a strong interest in *NTM* at a time when his own lunar theory was still very imperfect and incomplete. He compared his own results (i.e., **gors**) with those of others: first with Clairaut and Euler, leading to the soon-to-be-abandoned **griend-grond** variety of tables; and then with Newton's 'theory', which inspired him to the **geer** version modelled after Newton's theory. Several folios in Cod. $\mu_{28}^{\sharp}$ attest Mayer's engagement with Newton's lunar theory and his ability to translate its equations into the analytical language of trigonometric quantities. Although the dating of those folios is uncertain, their results are largely in accord with the list in Cod. $\mu_{15}^{\sharp}$, fol. 9r, in the column under the heading *New. Flamst.* There Mayer started a line of development characterized by the multistep scheme of application, with the middle step equivalent to the Horrocksian kinematics. Manuscript Cod. $\mu_{15}^{\sharp}$ in combination with the Mayer-Euler correspondence show how Mayer embraced the essence of the computational scheme of Newton's lunar theory during 1752. Mayer's remarks to Euler in a letter of January 1752, quoted above on p. 110, presage this development.

Mayer's change of strategy seems to have been made for pragmatic rather than theoretical reasons. It happened when his results remained inadequate,

when he was looking around for inspiration, and when he had shared with
Euler his intentions to employ a variable eccentricity as in Newton's theory.
The standard deviations listed in Display 6.3 show that the new plan, embodied in the *Entwurf neuer ☽ Tafeln*, brought about a dramatic fourfold
increase in accuracy. The reason for the surprising increase in accuracy after
adoption of Newton's scheme should be the subject of further research.

I am not aware of any position computations that immediately showed to
Mayer the success of his new scheme, nor of any indications that his change
was theoretically motivated. In the next chapter, we will see that all the multistepped computational schemes can be transformed into a single-stepped
scheme; this implies that – as far as accuracy is concerned – there is no real
advantage of any scheme over any other. In other words, the multistepped
scheme is not really necessary, eventually we will have to address the question
why Mayer adhered to it once he had the technical tools to get rid of it.

Mayer introduced the multistepped procedure in this stage of his work
as a consequence of his assimilation of Lemonnier's lunar tables, which were
rooted in Flamsteed's rendering of *NTM*, while *NTM* in turn incorporated
Horrocks's variable ellipse. Lemonnier's tables necessitated the user to perform quite involved additional interpolations, associated with the variable
eccentricity and the seasonal fluctuations of some equAtions. In contrast,
Mayer's tables were much more straightforward to apply. In his preface to
the kil tables, Mayer credited Euler for providing the means to make the
transition. We must understand his allusions to Euler not in connection with
a dynamical lunar theory, but rather in connection with the changing perception of trigonometry, which the latter had brought about. It is hard to imagine
that the computations in Sect. 6.4 would have been performed if every sine
and cosine were regarded as half a chord in a circle. Among the advances
in celestial mechanics taking place in the eighteenth century, the changing
perception of trigonometry may well have been as important a development
as the advent of the differential calculus.

The new *Entwurf* soon developed into the kil tables, printed in 1753. Mayer
was so proud of the latter that he boasted to Euler:

> So much is certain, that the tables give the Moon's position as accurately as could
> hitherto have been obtained through observations, and that therefore no greater
> reliability in these can be supplied or hoped for until more diligence is first of all
> applied to the method of observing.[51]

The accuracy of kil was widely recognized and highly appreciated; the tables were even accurate enough to bring the application of the lunar distance
method of longitude determination within reach. Thus, Mayer was encouraged to enter the quest for the Longitude Prize. Therefore, these tables play
a key role in Mayer's work on lunar motion. In their introductory comments,
Mayer showed that he had a firm understanding of the physical meaning of
each of its equAtions. Yet, contrasting with Mayer's claim, in the light of

---

[51] Mayer to Euler, 7 May 1753, Forbes (1971a, pp. 65–66).

their *NTM*-like way of application, it is unlikely that the kil tables of 1753 were backed by a coherent lunar theory on a dynamical basis. There may have been partial results derived from such a theory, e.g., his new equAtions of arguments $p - \varsigma$ and $2\omega - p - \varsigma$. Such results may perhaps provide the reason why Mayer was able to quickly improve upon Lemonnier's tables.

At least one contemporary astronomer might possibly have recognized the link between Mayer's kil tables and the Newton–Lemonnier theory, had he been searching for it. This was d'Alembert, who went to great lengths to compare his own lunar theory to Lemonnier's tables. He even preferred to present his own results in the format of that illustrous example. Kil reached d'Alembert while the last pages of his own theory were being printed, and he had just enough time left to wedge in a contemptuous comment on Mayer's tables. D'Alembert did make an effort to compare the works of several authors, including Mayer's, and he even recognized that Mayer's tables came closest to those of his own favourite. But apparently he did not recognize to what extent they both relied on the same source. Instead, he wondered whether Mayer had adopted *Euler's* theory, or whether he had a theory of his own.[52]

Another commentator that needs to be mentioned here is Petrus Frisius, who compared the theories of Newton, Clairaut, Euler, d'Alembert, and Mayer. He was able to convert the Horrocksian kinematic mechanism of Newton's *NTM* into trigonometric expressions and he also recognized that the position of evection had changed between Mayer's kil and rede tables. Yet he too was unable to discover the link between *NTM* and kil.[53] The form in which *NTM* was presented, be it in Newton's descriptive pamphlet or in Lemonnier's tables, was apparently just too different from Mayer's kil tables.

The final implication of all this seems to be that the influence of the differential calculus on the success of Mayer's lunar tables of 1753 is much less significant, and the impact of Newton's 1702 theory (more specifically, its implementation by Flamsteed as published by Lemonnier) much more significant, than has hitherto been assumed. The success of Mayer's 1753 lunar tables depends on a hybrid mix of these two ingredients, with a firm dash of coefficient fitting as a third component.

---

[52] d'Alembert (1754–1756, I pp. xxvi, 250–252). D'Alembert's quantitative statements do not all seem to make sense; For example, he asserted that Mayer's evection is effectively very different from Lemonnier's, while in truth both are almost the same after conversion into a comparable form. Later, d'Alembert recognized that Mayer's kil tables were the most accurate available and also the most practical to use, whereupon the French nobleman constructed new lunar tables with a multistep format (d'Alembert 1761–1780, II, p. 271–312).

[53] Frisius (1768, pp. 272–273, 357).

Chapter 7
Multisteps in *Theoria Lunae*

In Chap. 5 we have studied Mayer's lunar theory, and we have seen how Mayer had derived, from differential equations of motion and Newton's law of gravitation, a mathematical formula expressing the true longitude of the moon in terms of its mean motion. His solution of the differential equations amounts to what we have termed a single-stepped one. In contrast, we have seen in Chap. 4 that his lunar tables implemented a multistepped procedure, using not only mean motion arguments: several tables had to be entered with arguments that had been modified in the course of calculation. Chapter 6 demonstrated that the multistep procedure is rooted in Newton's *Theory of the Moon's Motion* (*NTM*) and that Mayer is unlikely to have had a theoretical justification of it. Thus, a gap shows up between theory and tables.

Yet, Mayer compiled *Theoria Lunae* in order to satisfy Bradley's interest in the theoretical background of the tables, and his declared intention with it was to show that from the Newtonian theory (i.e., the law of gravitation) no arguments against his tables could be drawn.[1] A gap between the single-stepped solution and the multistepped tables would certainly be regarded as such an argument. So the task awaited Mayer to reconcile the *mathematical* equation of the moon's motion, as derived from the differential equations, with the *astronomical* equAtions represented by the tables.

We will reconstruct Mayer's workaround in this chapter. First, we will undertake a detailed study of the relevant passages in *Theoria Lunae*. This will impart only Mayer's assertion that the theory and the tables could be linked, but it will fail to disclose how he accomplished the link. Moreover, his reasons for the multistepped scheme presented in *Theoria Lunae* will be shown to be insufficient in Sect. 7.2. After a closer investigation of exactly what needs to be clarified, we will present (Sect. 7.4) a candidate for Mayer's translation of the single-stepped solution into the multistepped form, applying a technique that is extant in some of his notebooks. The actual implementation of this

---

[1] See the preface of *Theoria Lunae* (Mayer 1767, p. not numbered), which is almost completely translated in Forbes and Wilson (1995, p. 65–66).

S.A. Wepster, *Between Theory and Observations*, Sources and Studies in the History of Mathematics and Physical Sciences, DOI 10.1007/978-1-4419-1314-2_7,
© Springer Science+Business Media, LLC 2010

technique (Sect. 7.5) leads us to a result well in accord with *Theoria Lunae*'s own multistepped solution, therefore we conclude in the end that Mayer indeed transformed the single-stepped solution into the multistepped form.

By way of caution, the text of *Theoria Lunae* presents the details of the multistep procedure not quite clearly, and Mayer explains the variables that are involved rather poorly. The *Theoria Lunae* text can only be understood in combination with a detailed understanding of the mode of operation of the multistep lunar tables. Section 4.1 may illuminate the reader who encounters difficulties.

7.1 *Theoria Lunae*'s Breakpoint

We have followed, in Chap. 5, Mayer's lunar theory up to the statement of its single-stepped longitude solution, reproduced in the left-hand column of Display 7.1 on p. 123, and aliased waard. We continue our discussion of *Theoria Lunae* at the point where we left off, in order to investigate how it introduced the multistep computational procedure. Three significant changes take place in one paragraph (§49) of *Theoria Lunae*. We will now investigate them in turn.

7.1.1 A Change in Procedure

The most noticeable change is indeed the replacement of the single-stepped format by a multistepped one. Right after having arrived at the single-step solution, Mayer wrote:

> Although this formula to find the true longitude of the moon φ from the mean q is apt enough and can easily be rendered in astronomical tables, in which the arguments of the inequalities are expressed by the mean motions of the anomaly, of the distance to the sun and of the distance to the node, in precisely the same way that the celebrated Clairaut arranged his tables; still the enormous number of inequalities, and consequently of constructed tables, would render the calculation of the place of the moon so very painstaking and annoying, that weariness could appear to even the most patient computer and deter him from examining the tables. Therefore I have looked for a way by which the inequalities that the formula contains can be reduced to a smaller number. And indeed, upon treatment of the whole matter from beginning to end, it appeared easily that many inequalities arise from the cause that the mean motion of the sun was introduced in the calculation, and that they would disappear if the true place of the sun was applied; next [it appeared] that the terms $-5171 \sin(\alpha + n)q$ and $+7615 \sin(\alpha - n)q$ are for the largest part taken away by applying the special equation that depends on the mean anomaly of the sun nq to the moon's mean anomaly αq, before the equation of centre is picked out by this last one. Finally [it appeared] that even other inequalities either diminish or completely disappear if to the same anomaly of the moon are applied the minor inequalities

Display 7.1 *Theoria Lunae* solutions, single-step left, multistep right (Mayer 1767, §§48, 49)

waard	wantij
$\varphi = q + 10^{-7}\,($	$\varphi = q + 10^{-7}\,($
$-1094159\,\sin\alpha q$	$-1096465\,\sin\tilde{p}$
$+\quad 37308\,\sin 2\alpha q$	$+\quad 37669\,\sin 2\tilde{p}$
$-\quad 1732\,\sin 3\alpha q$	$-\quad 1722\,\sin 3\tilde{p}$
$+\ 115046\,\sin 2rq$	$+\ 103850\,\sin 2\tilde{\omega}$
$+\quad\quad 600\,\sin 4rq$	$+\quad\quad 246\,\sin 4\tilde{\omega}$
$-\quad 9374\,\sin(2r+\alpha)q$	$+\quad 2665\,\sin(2\omega+p)$
$-\ 221514\,\sin(2r-\alpha)q$	$-\ 232298\,\sin(2\omega-p)$
$+\quad 1779\,\sin(4r-2\alpha)q$	$+\quad 1857\,\sin(4\omega-2p)$
$-\quad\quad 28\,\sin(4r+\alpha)q$	$+\quad\quad 122\,\sin(4\omega+p)$
$-\quad 1925\,\sin(4r-\alpha)q$	$+\quad\quad 441\,\sin(4\omega-p)$
$+\quad 1092\,\sin(2r+n)q$	$-\quad 2831\,\sin(2\omega+\varsigma)$
$-\quad 7969\,\sin(2r-n)q$	$-\quad 3534\,\sin(2\omega-\varsigma)$
$+\quad\quad 333\,\sin(2r+2\alpha)q$	$-\quad 1232\,\sin(2\omega+2p)$
$+\quad 9368\,\sin(2r-2\alpha)q$	$-\quad 3013\,\sin(2\omega-2p)$
$-\quad\quad 153\,\sin(4r-4\alpha)q$	$-\quad\quad 123\,\sin(4\omega-4p)$
$+\quad\quad 204\,\sin(2r-2n)q$	$-\quad\quad 26\,\sin(2\omega-2\varsigma)$
$-\quad\quad 43\,\sin(2r+\alpha+n)q$	$+\quad\quad 426\,\sin(2\omega+p+\varsigma)$
$-\quad 1188\,\sin(2r-\alpha+n)q$	$+\quad 6341\,\sin(2\omega-p+\varsigma)$
$+\quad\quad 610\,\sin(2r+\alpha-n)q$	$-\quad\quad 305\,\sin(2\omega+p-\varsigma)$
$+\quad 9992\,\sin(2r-\alpha-n)q$	$+\quad 2113\,\sin(2\omega-p-\varsigma)$
$+\quad 33983\,\sin nq$	$+\quad 33902\,\sin\varsigma$
$-\quad\quad 497\,\sin 2nq$	$-\quad\quad 497\,\sin 2\varsigma$
$-\quad 5171\,\sin(\alpha+n)q$	$+\quad\quad 566\,\sin(p+\varsigma)$
$+\quad 7615\,\sin(\alpha-n)q$	$+\quad 1942\,\sin(p-\varsigma)$
$-\quad\quad 282\,\sin(2r-3\alpha)q$	$+\quad 1093\,\sin(2\omega-3p)$
$+\quad\quad 549\,\sin(4r-3\alpha)q$	$+\quad\quad 585\,\sin(4\omega-3p)$
$-\quad 7702\,\sin rq$	$-\quad 7691\,\sin\tilde{\omega}$
$+\quad\quad 545\,\sin(r+\alpha)q$	$+\quad\quad 126\,\sin(\omega+p)$
$+\quad\quad 100\,\sin(r-\alpha)q$	$+\quad\quad 519\,\sin(\omega-p)$
$-\quad 1166\,\sin(r+n)q$	$-\quad 1024\,\sin(\omega+\varsigma)$
$+\quad\quad 295\,\sin(r-n)q$	$+\quad\quad 153\,\sin(\omega-\varsigma)$
$+\quad\quad 71\,\sin 3rq$	$+\quad\quad 71\,\sin 3\tilde{\omega}$
$-\quad 19931\,\sin 2iq$	$-\quad 19932\,\sin 2\tilde{\delta}$
$-\quad\quad 333\,\sin(2r+2i)q$	$-\quad\quad 102\,\sin(2\omega+2\delta)$
$+\quad 3933\,\sin(2r-2i)q$	$+\quad 4164\,\sin(2\omega-2\delta)$
$+\quad 2200\,\sin(2i+\alpha)q$	$+\quad\quad\quad 8\,\sin(2\delta+p)$
$+\quad 1433\,\sin(2i-\alpha)q$	$+\quad 3625\,\sin(2\tilde{\delta}-\tilde{p})$
$-\quad\quad 26\,\sin(2i+n)q$	$+\quad\quad\quad 9\,\sin(2\delta+\varsigma)$
$+\quad\quad\quad 5\,\sin(2i-n)q$	$-\quad\quad 30\,\sin(2\delta-\varsigma)$
$-\quad\quad 84\,\sin(2r-2i+\alpha)q$	$+\quad\quad 124\,\sin(2\omega-2\delta+p)$
$+\quad\quad 138\,\sin(2r\quad 2i-\alpha)q$	$-\quad\quad 76\,\sin(2\omega-2\delta-p)$
$-\quad\quad 33\,\sin(2r+2i-2\alpha)q$	$+\quad\quad\quad 3\,\sin(2\omega+2\delta-2p)$
$-\quad\quad 91\,\sin(2r-2i-2\alpha)q$	$-\quad\quad 73\,\sin(2\omega-2\delta-2p)$
$+\quad\quad 761\,\sin(2r-2i+n)q$	$+\quad\quad 619\,\sin(2\omega-2\delta+\varsigma)$
$-\quad\quad 605\,\sin(2r-2i-n)q$	$-\quad\quad 463\,\sin(2\omega-2\delta-\varsigma)$
$+\quad 1795\,\sin(2\alpha-2i)q)$	$+\quad 1795\,\sin(2p-2\delta))$

Coefficients are expressed in radian

of longitude, except some of them that depend on the distance of the moon to the sun rq and to the node iq.[2]

Clairaut's *Théorie de la Lune* contained an expression for the moon's longitude in 29 trigonometric terms, which could be cast into 22 tables each catering for one equAtion of a single argument, excluding the reduction to the ecliptic. This was considerably more than the 13 equAtions that Mayer had tabulated, and in that sense indeed 'weariness would appear' before one would have computed a single lunar position using Clairaut's results.[3] Apparently, Mayer knew Clairaut's theory when he wrote this passage but he certainly did not know it when he contrived the multistep procedure in 1752; at that time, Euler had forwarded Clairaut's coefficients to Mayer but not his theory.

In the quoted passage, Mayer clearly expressed that for practical reasons he wanted to reduce the number of terms that one had to compute or look up in tables. He achieved this goal in three ways. One way, he said, was to introduce the sun's true position instead of its mean position, since the latter was responsible for the presence of some (unspecified) terms. Apparently, with 'true position' of the sun he meant its true longitude, but not its true anomaly, since he kept the mean anomaly instead of the latter.[4] Thus, this change would affect terms that have the lunar elongation in their argument, but not the terms that depend on solar anomaly only.[5]

Further, he had found that certain terms are greatly reduced when the moon's mean anomaly is first corrected by what he called a 'special equAtion'.

---

[2] '*Quanquam haec formula ad longitudinem Lunae veram φ ex media q inveniendam satis apta sit atque in tabulas astronomicas facile redigi possit, in quibus argumenta inaequalitatum per motus medios anomaliae, distantiae a Sole et nodo exprimantur, eo prorsus modo, quo suas disposuit tabulas celeb[er] Clairaut; ingens tamen numerus inaequalitatum, et tabularum inde construendam calculum loci Lunae adeo operosum ac molestum redderet, ut vel patientissimo calculatori taedium parere eumque ab examine tabularum deterrere possit. Idcirco viam quaesivi, qua ad pauciores redigi queant inaequalitates, quas ista formula complectitur, et quidem tractanti hanc rem universam ab initio ad finem facile adparebat, plures inaequalitates inde ortas esse, quod in calculum medius motus Solis introductus sit, quae disparae erant, si verus Solis locus fuisset adhibitus; deinde terminos $-5171\sin(\alpha+n)q$ et $+7615\sin(\alpha-n)q$ maximam partem tolli adplicando anomaliae mediae Lunae αq, antequam per hanc aequatio centri excerpatur, aequationem peculiarem, quae ab anomalia media Solis nq pendeat. Denique etiam alias inaequalitates vel minui, vel evanescere prorsus, si eidem anomaliae Lunae adplicentur inaequalitates longitudinis minores, excepta parte earum, quae a distantia Lunae a Sole rq et a nodo iq pendent*' (Mayer 1767, pp. 47–48), partly quoted in Forbes and Wilson (1995, p. 64).

[3] See Clairaut (1752b, pp. 54, 89).

[4] This is apparent from, for instance, the kil and rede tables, and our example calculation on page 52. Also see letter to Euler (Forbes 1971a, p. 61).

[5] D'Alembert discussed the effect of the true vs. mean sun on the lunar tables of Lemonnier (d'Alembert 1754–1756, III, pp. 47–49). He envisaged that the variation and the equAtions of arguments $2\omega \pm s$ and $2\omega - p \pm s$ would be affected. This part of d'Alembert's work was published in 1756, i.e., after Mayer had completed *Theoria Lunae*, so it was not known to Mayer at the time.

The engaged terms were those with arguments $(\alpha \pm n)q$, i.e., the sum or difference of the mean anomalies of sun and moon. It turns out that the 'special equAtion' is $+22'6'' \sin \varsigma - 15'' \sin 2\varsigma$; this is the annual equAtion of anomaly, which Newton had introduced (see p. 97). This equAtion is closely related to the annual equAtion of the node, likewise discovered by Newton, and likewise incorporated in Mayer's tables, but not mentioned here in his *Theoria Lunae* text.

Additionally, Mayer had found that the number and magnitude of terms decreased further when some of them (which he called the 'minor inequA-lities') were used to correct the anomaly. Thus, the 'special equAtion' and 'minor equAtions' had to be applied to the mean lunar arguments, yielding an intermediate anomaly with which to compute the lunar equAtion of centre.

In short, we see that this is the place where Mayer introduced his multistep procedure into *Theoria Lunae*.

The quoted passage is quite hard to understand without an understanding of the multistepped format embodied in the tables, and that is precisely why an example calculation has been included in Sect. 4.1 above. It is not clear from the text alone which equAtions were exactly to be subsumed under these 'minor inequAlities' (they will turn out to be those numbered from I to X in the lunar tables as they occur on page 52). Mayer's Latin is more complicated in this passage than usually, too. Usually, he was able to explain matters very clearly, but in this passage he remained rather vague. Only someone with excellent knowledge of his tables would be able to understand him here. At this point it should be reminded that *Theoria Lunae* was written at the request of James Bradley, who showed that he was well acquainted with the published kil tables as well as with the rak manuscript tables, which Mayer had sent in the winter of 1754–1755. Mayer could expect that at least his prime reader had excellent knowledge of the tables; yet this passage makes a somewhat cumbersome impression.

7.1.2 A Change in Notation

In the same paragraph of *Theoria Lunae*, and immediately following the passage quoted above, Mayer introduced a major change in notation. Up till here, he had expressed the arguments of the equAtions as follows:

αq lunar mean anomaly
nq solar mean anomaly
rq mean elongation
iq mean nodal distance of the moon

In this notation, all arguments were expressed as products of the time-like variable q and a mean motion constant, so they represent mean motions. But

right after the above-mentioned change in procedure, Mayer introduced the
following new variables:[6]

p	lunar mean anomaly (actually in Mayer's text $\dot{p}$)
$\tilde{p}$	lunar anomaly corrected by the 'minor equAtions'
ς	solar mean anomaly
ω	elongation of the mean moon from the true sun
$\tilde{\omega}$	elongation of the 'equAted' moon from the true sun
δ	distance of the mean moon from the 'corrected' node
$\tilde{\delta}$	distance of the 'equAted', and also for variation corrected, moon from the 'corrected' node

The words in quotes are not fully explained in his text and they are not
straightforward to understand for a modern reader. It is not difficult to understand that 'equAted' means 'after applying one or more equAtions', but it
is impossible to infer from this or the preceding text *which* equAtions were
involved. From the text it is also impossible to make out what the 'minor
equAtions' are, and what the difference is between 'equAted' and 'corrected'.
We will return to these questions in Sect. 7.3. Perhaps one or more of these
uncertainties were clear to a contemporary astronomer.

The dot on $\dot{p}$ is Mayer's way to distinguish his usual identifier for lunar
anomaly from the anomaly-like independent variable of his preceding theory.
At that time the British still favoured the dot-notation for differentiation
while on the continent the Leibnizian d-ism was preferred. Therefore, Mayer's
use of a dot by way of accent has a tinge of humour in it, when we realize that
he, a continental, wrote *Theoria Lunae* on the explicit request of the British
Astronomer Royal. Mayer had used essentially the same symbols (p, ς, ω,
and δ; without accents) already years before *Theoria Lunae*. They appeared
in his manuscripts and in several letters of Mayer's to Euler.[7] Since the danger
of confusing the two uses of the identifier p is minimal, we will drop the
dot and write p for the lunar mean anomaly. Also, the Greek letter ς is
indistinguishable from the Roman letter s in Mayer's handwriting. We adhere
to Maskelyne's reading of the *Theoria Lunae* manuscript and, for consistency,
we always use ς for solar mean anomaly.

Mayer's former notation, in which all the arguments were multiples of q,
implied that at the moment in history when $q = 0$, all arguments were 0, i.e.,
the *true* sun and moon were both in their own apogee, while both apogees
coincided with the ascending node of the lunar orbit. This implies an annular
solar eclipse under rather special circumstances.The new symbols that he
introduced here, suggest a greater independence of the arguments from each
other. The special conjunction at $q = 0$ is no longer implied.

---

[6] Mayer (1767, p. 48).

[7] See the letters of 4 July 1751 and 7 January 1753 in Forbes (1971a, p. 63).

7.1.3 A Change in Word Meaning

Third, the word 'equation' underwent a change in meaning in this paragraph of *Theoria Lunae*. Whenever Mayer used the word 'equation' in the preceding text of *Theoria Lunae*, it was always in the sense of a mathematical equation, but in this paragraph he used it in the astronomical sense. Thereafter the word 'equation' did not reappear until the separate treatment of the equAtion for the lunar latitude (i.e., (III′) in Chap. 5). Instead, he often used the word 'inequality', which is for us maybe even more troublesome than the word 'equation'. This word too has a different connotation for astronomers and mathematicians, so we have to distinguish the well-known mathematical inequality from the astronomical inequAlity, which is an irregularity giving rise to an equAtion. In *Theoria Lunae*, Mayer used the word 'inequality' only sparingly in the astronomical sense, and he does not employ it in its mathematical meaning.

In view of the three changes just discussed, we must conclude that this §49 of Mayer's is a turning point in *Theoria Lunae*: the single-step solution is somehow replaced by the more complicated multistepped scheme, the notation for the arguments changes drastically, and his language turns into that of an astronomer rather than a mathematician. Up to this point Mayer had put emphasis on the theoretical derivation of a mathematical equation for the longitude of the moon, but here he makes a turn towards astronomical tables. This is further exemplified by his change from radians to degrees as the unit to express the coefficients of the terms: radians were natural to use when he substituted trigonometric series in the differential equations of motion (see Chap. 5), but degrees were desirable for the construction of tables for regular astronomical use.

7.2 Failing Assertions

Forbes and Wilson were without doubt referring to the text we have been discussing above, when they remarked:

> In his *Theoria Lunae* Mayer showed that if the terms derived from theory, and depending solely on mean arguments, were transformed so as to depend on the progressively modified arguments [...] then many of the resulting terms were so small as to be safely neglected. This theoretical justification, however, was not made public [i.e., in *Theoria Lunae*] till long after the tables of 1753 were published.[8]

This is a fitting summary, except for one word: 'justification'. Besides, the theory attempted to clarify not the kil tables of 1753 but the later rak tables with their revised position of evection.

---

[8] Forbes and Wilson (1995, p. 65).

Mayer did not justify much in this part of his treatise. We have just seen him proposing some changes, with the specific goal of reducing the necessary number of equAtions, but an explanation why these changes would achieve the stated goal was lacking. He asserted that the original single-stepped solution waard (Display 7.1, lefthand column) took on the new form wantij (righthand column) after implementing the aforementioned changes. In modern terms, his assertion amounts to a mathematical transformation between the two. There is more at stake than a mere renaming of arguments, such as $\alpha q \to p$, $nq \to \varsigma$, $rq \to \omega$, and $iq \to \delta$, because the transformation implies changes to the coefficients that reflect the progressive changes in the arguments. We will investigate this transformation in the following sections of this chapter.

Were Mayer's claims of reducing the size and number of the equAtions substantiated? Indeed, we observe that the terms involving the two arguments $(\alpha \pm n)q$ are greatly reduced after the transformation, when they reappear as $p \pm \varsigma$. There are other terms, too, whose magnitude is reduced, but also some that have increased, and quite some have changed sign. I would not conclude that in general a spectacular decrease in the magnitude of the terms has been accomplished, or even that there are now many more negligible terms. When selecting the equAtions to be tabulated, Mayer apparently neglected terms smaller than about $30'' \approx 0.00015\,\mathrm{rad}$ (except when they could be combined with larger terms in one equAtion, e.g., the annual equAtion combines $11'46'' \sin s$ with $10'' \sin 2s$).

Figure 7.1 shows a plot of the coefficients before and after the transform. For each term occurring in the series solution of the lunar longitude, a pair of coordinates is formed from the absolute values of the coefficients in Display 7.1, taking the coefficient before the transform for the horizontal coordinate, and the coefficient of the corresponding term after the transform for the vertical coordinate. Thus, an only slightly affected term will be situated near the main diagonal of the plot. Likewise, a term that is reduced by the transform will fall below the diagonal. The plot area is divided into four rectangles by lines at Mayer's cutoff magnitude of $30''$.

The figure shows immediately that the terms that get reduced by the transform are nearly balanced by ones that get magnified, while many are not significantly affected at all. Most of the would-be significant changes occur in the lower left square, signifying they were already below the cutoff before the transform and stayed there. In the lower right rectangle, where we find those coefficients that are smaller than the cutoff after the transformation but not before, three dots stand out clearly; also three dots are in the reverse situation in the upper left rectangle. The five dots below the main diagonal in the upper right region signify coefficients that *are* significantly reduced, but not so much that their terms may be neglected.

We must conclude that Mayer's claim was not substantiated: the multistep transform reduced neither the number of significant terms nor the number of equAtions, contradicting his stated reason for its implementation in *Theoria Lunae*.

Fig. 7.1 Log–log plot of absolute values of coefficients. The horizontal axis represents magnitudes before the transform, and the vertical axis after the transform. Mayer's cutoff amplitude of $30'' \approx$ 0.00015 rad is indicated by the *grid lines*

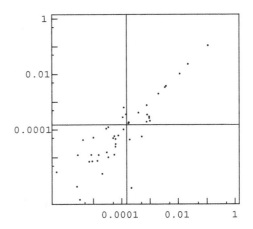

Besides, we note the puzzlingly late replacement of mean solar longitude by true solar longitude. At this late stage in his treatise, after all the long and tedious calculations to derive the single-step solution (as expounded in Chap. 5), Mayer confessed that he had started out at the beginning with an argument that was unsuitable. If he redid his calculations using the true longitude, as he tells us, why then did he not include this more useful version in the manuscript? Sure enough the neglect was not a result of discovering his mistake only when the manuscript was almost complete: writing *Theoria Lunae* in 1755, he had already decided in favour of the solar true longitude at least two years earlier.[9] Even if he had a complete lunar theory lying on the shelf at that time, then still the competition for the Longitude Prize would be a first-class occasion to bring it up to standard.

In conclusion, Mayer did not explain to his reader *why* the multistep solution should be any better than the single-step one. He revealed only one reason for the multistep scheme, namely that it reduced the magnitude of several equAtions – a claim that we have just seen to be not justified. Besides, the reader of *Theoria Lunae* is left wondering why Mayer had chosen just this particular scheme. The answer is partly contained in the disclosures of the previous chapter: Mayer developed the multistep procedure from Newton's *NTM* and Lemonnier's lunar tables, and it brought him a fabulous improvement in accuracy. When he was writing *Theoria Lunae*, the lunar tables with their multistepped design were already in Bradley's hands. He just had to account for it either one way or the other. Mayer held a firm belief, fostered by experience, that the multistepped scheme was more accurate. But he could not account theoretically for his experience, a vexation which he obscured in the passage here scrutinized.

---

[9] Mayer to Euler, 7 January 1753, (Forbes 1971a, p. 61). Euler had recommended the true motions already in 1751 (Forbes 1971a, pp. 37–38).

We understand that here lay a particularly difficult point for Mayer to resolve. *Theoria Lunae* was meant to supply a theoretical foundation to tables that had evolved out of a mainly pragmatic conduct. Mayer had expressed his aim with *Theoria Lunae* in its preface: his goal was not to derive the tables completely from theory, but to show that the theory could provide no argument against his tables. The gap between theory and tables had to be bridged convincingly for Bradley, or else the Longitude Prize would be in danger. This predicament may have caused a substantial part of the delay in producing *Theoria Lunae*.

7.3 Substitution of Arguments

Mayer's assertion about the benefit of the multistep scheme now disproved, we turn to the question that will occupy us for the rest of this chapter: did Mayer tranform *Theoria Lunae*'s single-step solution into *Theoria Lunae*'s multistepped solution, and if so: how did he do it? In other words, we have to investigate how Mayer may have transformed the single-stepped waard column of Display 7.1 into the multistepped wantij column of the same display by substitution of the new arguments discussed in Sect. 7.1.2.

We will proceed as follows. First, we need a clearer view of what needs to be transformed into what. This is provided in the current section. Then in Sect. 7.4 will be presented a technique, which I discovered among Mayer's manuscripts, that is essential to get the end result in the required multi-stepped form. Next, we actually transform the equation of the single-step solution, and we compare the result with Mayer's multistep equation. The two will be seen to be so similar that we have no reason to doubt whether Mayer really transformed the single-step to the multistep form in *Theoria Lunae*.

What, then, are the substitutions by which the single-stepped solution is transformed into the multistep scheme? This question is impossible to answer from the *Theoria Lunae* text alone, as we have seen in Sect. 7.1. We may gather additional information from the transformed equation of lunar longitude, i.e., wantij in Display 7.1, and from Mayer's post-1754 lunar tables of the rak and rede varieties. We will use Mayer's new arguments as introduced above in Sect. 7.1.2. I derived their meanings, which Mayer did not explain very clearly, mainly from the rede tables, which were also applied in the example calculation in Sect. 4.1. Now follows a description of the actions to be taken to replace the arguments of the single-step solution by the arguments of the multistep solution. The algorithm-like description will be useful in the sequel when we set up the transform between the single-stepped and multistepped forms.

1. Rename the mean anomaly of the sun nq to ς, and the mean anomaly of the moon αq to p.

These arguments are merely renamed, their values remain unaltered:

$$\varsigma = nq,$$
$$p = \alpha q.$$

Recall that p now denotes the mean anomaly of the moon, not any longer the independent variable used in the derivation of the single-stepped solution.

2. Substitute the longitude of the true sun in place of the mean sun.

In the quoted text of *Theoria Lunae*, Mayer suggested that he had redone all his calculations,[10] which he had summarized on the foregoing 47 pages of *Theoria Lunae*. Instead of reworking the complete theory of Chap. 5 now with the true sun, we will take a shortcut and apply the solar equAtion of centre to the longitude of the mean sun, wherever it appears in the single-stepped equation. Because Mayer maintained the solar mean anomaly, the only affected argument is the elongation rq, for which we substitute ω using

$$\omega = rq + 2e\sin\varsigma - \tfrac{5}{4}e^2 \sin 2\varsigma. \tag{7.1}$$

Here e represents the eccentricity of the sun's orbit. *Theoria Lunae*'s value for e is[11] 0.0168350. Note that angles must be expressed in radians in this formula.

3. Compute the 'minor inequAlities of longitude, except some of them that depend on the distance of the moon to the sun [rq or ω] and to the node' [iq or δ].[12]

Which are these minor inequAlities, i.e., minor equAtions? This is not clear from *Theoria Lunae*'s text. The difficulty is the interpretation of the words 'some' and 'depend'. Clearly we have to exclude terms that have argument $k\omega$ or $k\delta$ for integer values of k, but should we also exclude terms depending on mixed arguments, such as $k\omega \pm p$? We know that the lunar tables included such terms among the minor equations, so most likely they are meant to be in. On the other hand, the equAtion of centre is obviously not a minor one, so the terms depending on arguments kp are not among the minor equAtions, which also agrees with the tables.

So in principle *all* the equAtions that do not solely depend on kp, $k\omega$, or $k\delta$ should be considered as minor equAtions. However, to do so would

---

[10] 'treating the whole matter from beginning to end', (Mayer 1767, p. 48); as in the translation on page 122 above.

[11] Mayer (1767, p. 33). As explained in Sect. 1.3, it has long been conventional to measure solar anomaly from the *aphelion*; therefore, Mayer's anomaly ς differs from the modern anomaly by 180°. The change in convention is reflected in a sign change in the first-order term $2e\sin\varsigma$ of the equAtion of centre, but not in the second-order term. Therefore, the corrected elongation, which is found by subtracting the sun's true longitude from the moon's mean longitude, comes out as stated in the formula.

[12] Mayer (1767, p. 48); see also p. 122 above.

imply an enormous amount of useless work. A glance ahead at the next item in our list will show that the minor equations primarily serve to equate the lunar anomaly p. The anomaly is used only *indirectly*, i.e., as an argument for subsequent equations, for the computation of lunar longitude.[13] Therefore, a small error in the value of the anomaly will have only a small effect on the value of equations that depend on it. By far the largest of these is the first term in the equation of centre, $0.109 \sin p$; an error Δp in p would introduce an error of about $0.1 \Delta p \cos p$ in the moon's computed longitude. Therefore, any term that is neglected when equating p introduces an error in longitude of about one tenth of its magnitude. Thus it is sufficient to consider only the terms with the largest coefficients, say down to $15 \cdot 10^{-5} \approx 30''$. Doing so, we neglect terms that would contribute $3''$ or less to the final longitude. Therefore, we can ignore all terms except those depending on the following arguments:

$$2\omega - p,\ 4\omega - 2p,\ \varsigma,\ 2\omega - p - \varsigma,\ 2\omega - p + \varsigma,\ 2\omega + p,\ 2\omega - 2p,$$
$$2\omega - \varsigma,\ 2\omega + \varsigma,\ p - \varsigma,\ p + \varsigma,\ 2\delta + p,\ 2\delta - 2p,\ 2\omega - 2\delta,\ \text{and}\ 4\omega - p. \quad (7.2)$$

In the sequel, we will use the symbol $\mathcal{M}$ for the sum of the minor equations depending on these arguments.

4. Apply the annual equation of the anomaly and the minor equations to the lunar mean anomaly.

This annual equation is what Mayer called the 'special equation' of anomaly. *Theoria Lunae* states that this equation is $22'6'' \sin \varsigma - 15'' \sin 2\varsigma$; with this stated in radians we get for the corrected lunar anomaly:

$$\tilde{p} = p + 0.0064286 \sin \varsigma - 0.0000727 \sin 2\varsigma + \mathcal{M}. \quad (7.3)$$

5. Apply the annual equation of the node.

The tables usually include an annual equation of the node, but Mayer does not explicitly mention it in *Theoria Lunae*. It is implied, however, by his definition of δ as the distance of the mean moon from the *corrected* node. From the example calculation on page 53, it is evident that the annual equation is the primary correction to be applied to the node. This equation is approximately minus half of the annual equation of anomaly. For lack of a better value we let ourselves be inspired by (7.3) and take

$$\delta = iq - 0.0032143 \sin \varsigma + 0.0000364 \sin 2\varsigma.$$

6. Equate ω and δ, to get $\tilde{\omega}$ and $\tilde{\delta}$.

---

[13] Of course the 'smaller minor equations' do have a pronounced effect on lunar longitude directly, because the minor equations not only equate the anomaly, but also the longitude. However, this need not concern us at the moment, because here we strive to understand the relation between the mean anomaly p and its 'corrected' value $\tilde{p}$.

Again, the computational procedure of the lunar tables helps us to understand how to equAte in this case: the elongation of the moon is equAted by the minor equAtions and the equation of centre (argument $\tilde{p}$). The nodal distance δ is equAted by all equAtions up to and including the variation (argument $\tilde{\omega}$). If we write the lunar equAtion of centre as $\mathcal{C} = \sum_{k=1}^{4} c_k \sin k\tilde{p}$ and the variation as $\mathcal{V} = \sum_{k=1}^{4} v_k \sin k\tilde{\omega}$ with coefficients c_k, v_k not known before this particular stage of the transformation, then we have:

$$\tilde{\omega} = \omega + \mathcal{M} + \mathcal{C}, \tag{7.4}$$
$$\tilde{\delta} = \delta + \mathcal{M} + \mathcal{C} + \mathcal{V}.$$

This completes our list relating the old to the new arguments. Before we proceed, we make a note with respect to the place of evection (with argument $2\omega - p$), which we have here daringly included among the minor equAtions in $\mathcal{M}$. As we have seen on page 55, in lunar tables of Mayer prior to November 1754, the evection was not one of the minor equAtions. Instead, the evection was then applied simultaneously with the equAtion of centre. In those earlier formats, the argument of evection was equAted by the minor equAtions only, and the argument of the next equAtion, the variation, was equAted by both the equAtion of centre and evection. Which of the two schemes did Mayer adhere to in *Theoria Lunae*? In his multistep equation, the evection argument is written $2\omega - p$, not $2\tilde{\omega} - \tilde{p}$, suggesting a non-equAted argument. This would imply that, in *Theoria Lunae*, evection had its place among the minor equAtions. We will have an independent verification of this later in the chapter.

7.4 An Approximation Technique

In the previous section, we have unravelled the meaning of the new arguments that Mayer introduced in his *Theoria Lunae* text. But we are not yet ready to translate the lefthand side of Display 7.1 into its righthand side. To see why, consider the following.

Equation (7.1) teaches us the meaning of the new argument ω. To use this new argument, we have to substitute the 'old' argument rq by

$$rq = \omega - 2e \sin \varsigma + \tfrac{5}{4} e^2 \sin 2\varsigma,$$

wherever it appears; in like manner we have to substitute all other arguments. But by doing so we see that we run into an objectionable situation. Consider what happens to the term $0.0007702 \sin(rq)$ of the single-step solution (it is just over half way down the list in Display 7.1) when we substitute for rq:

$$0.0007702 \sin(rq) = 0.0007702 \sin(\omega - 2e \sin \varsigma + \tfrac{5}{4} e^2 \sin 2\varsigma). \tag{7.5}$$

The result is clearly not of a form we desire because the outer sine is a nonlinear function of two arguments. We would like to express the right-hand side as a sum of sines of linear combinations of arguments, so that the formula may be easily rendered into tables. The transformed solution in the wantij column of Display 7.1 shows the forms we want to obtain. So what we need is a technique to reduce the complicated forms to the simpler forms suitable for tables. Mayer was in dire need of such a technique, too.

And he had exactly that: I discovered it in one of his notebooks, under the heading *Reductio tabb mearum* ☽ *ad motus medios*.[14] These notebook pages contain the reduction of multistep equAtions to their equivalent single-step form. The goal of his exercise in the notebook was to compare these equAtions with the lunar theories of Clairaut and d'Alembert, as well as with another theory of himself. He had labelled the latter *juxta calc. theor. corr.*[15] and it is apparently almost the same as the single-step solution in *Theoria Lunae* (waard, reproduced in Display 7.1). The manuscript pages contain the reverse calculation of what we are after in this chapter, but the technique is the same. Because the technique is general, it deserves to be treated in more general symbols for arguments and coefficients than just the symbols in *Theoria Lunae*.

Suppose that x, x', and y denote arguments, and that η is a small coefficient, typically 10^{-3} or 10^{-4} rad. Let x' be x equAted by $\eta \sin y$, thus:

$$x' = x + \eta \sin y.$$

We endeavour to express $\sin x'$ as a sum of sine functions of simple linear combinations of x and y. Using well-known tigonometric realtions and Taylor approximations we have:

$$
\begin{aligned}
\sin x' &= \sin x \cos(\eta \sin y) + \cos x \sin(\eta \sin y) \\
&= \sin x (1 + \mathcal{O}(\eta^2)) + \cos x (\eta \sin y + \mathcal{O}(\eta^3)) \\
&= \sin x + \eta \cos x \sin y + \mathcal{O}(\eta^2) \\
&= \sin x + \tfrac{1}{2}\eta \left(\sin(x + y) - \sin(x - y) \right) + \mathcal{O}(\eta^2) \qquad (\eta \to 0).
\end{aligned}
\tag{7.6}
$$

So to the first order in η we have:

$$\sin x' \approx \sin x + \tfrac{1}{2}\eta \sin(x + y) - \tfrac{1}{2}\eta \sin(x - y), \tag{7.7}$$

which is an equation that matches the required form. When we extend the approximation to second order in η we find that the next terms are

$$-\tfrac{1}{2}\eta^2 \sin x \sin^2 y = -\tfrac{1}{4}\eta^2 \sin x + \tfrac{1}{8}\eta^2 \sin(x - 2y) + \tfrac{1}{8}\eta^2 \sin(x + 2y). \tag{7.8}$$

---

[14] 'Reduction of my lunar tables to mean motions', fol. 2r–7r in Cod. μ_8. With hindsight the same technique can be found in several other places in the manuscript.

[15] This may be interpreted as 'According to calculation by the corrected theory', or equivalently 'According to corrected theoretical calculation'.

In only a few isolated cases will we need these second order terms. Usually η is small enough to neglect them, especially since $\sin x'$ will itself be a constituent of an equation, where it will be multiplied by a similarly small coefficient. For the moment we leave the second order terms for what they are and concentrate on the linear terms.

Referring to our introductory example in (7.5), we note that a linear approximation of a more general form would come in handy. This can easily be obtained as follows. Let $x'' = x + \sum_{i=1}^{n} \eta_i \sin(y_i)$, for fixed n, with arguments y_i and small coefficients η_i. From (7.6) and induction it follows that

$$\sin x'' = \sin x + \tfrac{1}{2} \sum_{i=1}^{n} \eta_i \sin(x + y_i) - \tfrac{1}{2} \sum_{i=1}^{n} \eta_i \sin(x - y_i) + \mathcal{O}(\hat{\eta}_n^2), \qquad (\hat{\eta} \to 0)$$

(7.9)

where $\hat{\eta}_n = \max_{i=1}^{n} |\eta_i|$. When the second-order terms are neglected, the linear approximation results. Note that for $n = 1$ this is identical to (7.6).

As already mentioned, Mayer used this technique to reduce the equations underlying his multistepped tables to equivalent single-stepped form. These tables, which were of the **rede** variety and incorporated the latest improvements from continued efforts of fitting by Mayer, certainly date from later than 1755. We now conjecture that he used the same technique in the reverse direction, to derive the coefficients of the multistepped form out of the single-stepped form, when composing *Theoria Lunae* during 1755. Judging by the manuscripts, it seems unlikely that Mayer possessed this ability before the fall of 1754. D'Alembert was familiar with it, as appears from his *Recherches sur differens points importants du système du monde*, where he reduced Lemonnier's and Mayer's lunar tables to single-stepped form, in order to facilitate a comparison between the theories and tables of himself, Clairaut, Lemonnier, and Mayer.[16] As the first volume of d'Alembert's work appeared in 1754, it may be that Mayer acquired his knowledge through that source. Interestingly, J. K. Schulze reduced Mayer's tables to their single-step form by using elaborate Taylor expansions; application of the short-cut provided by (7.9) would have shortened his memoir considerably.[17] Apparently, then, this shortcut was no commonplace technique in the eighteenth century, perhaps because of its dedication to a very specific task.

---

[16] D'Alembert's reduction is only partial, since he considered only the major effects on the coefficients through the equation of centre, and he did not distinguish the individual terms of some composed equations, such as the three or four terms of the equation of centre and the two terms of the evection (d'Alembert 1754–1756, I, pp. 80, 89–95; and III, pp. 24–28). Laplace, in his turn, reduced the Mayer-based lunar tables of Mason and Bürg to a form comparable to his own theory, i.e., with true lunar longitude as the independent variable. Laplace employed first or second order approximations as circumstances dictated (Laplace 1802, pp. 645–651).

[17] In Schulze (1781), he reduced the multistep equations underlying **rede**.

7.5 The Complete Transform

Now it is time to put together the knowledge that we gathered in the previous sections, to reconstruct the transform of the single-stepped waard into the multistepped wantij (cf. Display 7.1). All we have to do is substitute the arguments as prescribed in Sect. 7.3, and expand the resulting terms with sine-dependent arguments using the approximation of Sect. 7.4. The reader is encouraged to check whether the steps, as implemented in this reconstructed transform, are in agreement with the example calculation in Sect. 4.4.

First, using the results of Sect. 7.3, we substitute the new arguments:[18]

$$nq = \varsigma, \qquad \alpha q = p, \qquad iq = \delta, \qquad \text{and}$$
$$rq = \omega - 0.0336700 \sin\varsigma + 0.0003543 \sin 2\varsigma.$$

We expand the terms where ω appears in the argument with the help of (7.9). For example in the term with argument $(2r - 2\alpha)q$, the 14th from the top of Display 7.1, the new arguments are introduced as follows:

$$0.0009368 \sin(2rq - 2\alpha q)$$
$$= 0.0009368 \sin(2\omega - 2p - 0.0673400 \sin\varsigma + 0.0007086 \sin 2\varsigma)$$
$$= 0.0009368 \sin(2\omega - 2p)$$
$$\quad - 0.0000315 \sin(2\omega - 2p + \varsigma) + 0.0000007 \sin(2\omega - 2p + 2\varsigma)$$
$$\quad + 0.0000315 \sin(2\omega - 2p - \varsigma) - 0.0000007 \sin(2\omega - 2p - 2\varsigma).$$

After substituting all the new arguments, and expanding where applicable, we select the minor equAtions specified by (7.2); let $\mathcal{M}$ be their sum. Then we transform the equAtion of centre, using (7.3) to derive the substitution

$$p = \tilde{p} - \mathcal{M} - 0.0064286 \sin\varsigma + 0.0000727 \sin 2\varsigma, \qquad (7.10)$$

which we insert in all terms with arguments kp, $k \in \{1, 2, 3, 4\}$. Where applicable we expand again using (7.9).

The new variation is found in a similar way, by the substitution in all terms with arguments $k\omega$, using (7.4):

$$\omega = \tilde{\omega} - \mathcal{M} - \mathcal{C},$$

followed by expansion. Finally, we substitute and expand the terms with arguments 2δ and $2\delta - p$, remembering that the annual equAtion of the node must be applied to δ. The former argument becomes:

$$2\delta = 2(\tilde{\delta} + 0.0026810 \sin\varsigma - 0.0000267 \sin 2\varsigma - \mathcal{M} - \mathcal{C} - \mathcal{V}) \qquad (7.11)$$

---

[18] These substitutions may look strange, but then remember that they take place in the waard column of Display 7.1.

and the latter then follows from (7.11) and (7.10):

$$2\delta - p = 2\tilde{\delta} - \tilde{p} + 0.0091096 \sin\varsigma - 0.0000994 \sin 2\varsigma - \mathcal{M} - 2\mathcal{C} - 2\mathcal{V}.$$

Now let us pay attention to second-order terms. Consider (7.7), and note that $\sin x'$ is multiplied by a coefficient, say ξ, because of it being part of an equation. Combining the first and second orders of (7.7) and (7.8), we observe that the full second-order expansion is

$$\xi \sin x' \approx \xi(1 - \tfrac{1}{4}\eta^2) \sin x \pm \tfrac{1}{2}\xi\eta \sin(x \pm y) + \tfrac{1}{8}\xi\eta^2 \sin(x \pm 2y).$$

Mayer wrote down seven decimal places in the equations before and after the transformation: at that precision, second-order expansions would have been necessary whenever $\tfrac{1}{4}\xi\eta^2 > 10^{-7}$. However, numerical experiments (to be discussed in the next section) suggest that the actual accuracy of his calculations is limited to five decimal places, or about $2''$. The second-order terms with the largest coefficients are those with largest ξ and η, which we find in the substitution of the evection (here one of the minor equations) in the equation of centre, and again in the substitution of the equation of centre in the variation. The evection yields the following second-order terms in the equation of centre:

$$-0.0000135 \sin p \approx -3'' \sin p$$
$$+0.0000067 \sin(4\omega - p) \approx +1'' \sin(4\omega - p)$$
$$-0.0000067 \sin(4\omega - 3p) \approx -1'' \sin(4\omega - 3p),$$

and the equation of centre in turn affects the variation as:

$$+0.0001229 \sin 2\tilde{\omega} \approx +25'' \sin 2\tilde{\omega}, \qquad (7.12)$$
$$-0.0000614 \sin(2\omega \pm 2p) \approx -13'' \sin(2\omega \pm 2p), \qquad (7.13)$$

neglecting the difference between arguments with and without tildes. Apparently, the second-order terms are significant only in the last substitution, in all other cases they contribute at most a few arcseconds and hence they are negligible.

I implemented the above interpretation of the transform in the computer program *Mathematica*, including the approximations to first-order, and where necessary to second-order. Display 7.2 lists the coefficients computed by the program, and the differences with respect to Mayer's values. In the next section, we will discuss the successes and failures of our interpretation.

Display 7.2 The multistep transform compared with Mayer's result. Columns are: argument; our computed coefficient (in 10^{-7} rad); difference with Mayer's value in 10^{-7} rad; idem in 10^{-7} rad; idem in ″. Terms absent from Mayer's list are taken as 0 there, and marked by an asterisk

Argument	Coefficient	Diff.	″
$2\omega - 2p - \varsigma$	+469	+469	+10*
$2\omega - p + 2\delta$	−452	−452	−9*
$2p + \varsigma$	−391	−391	−8*
$2p - \varsigma$	+391	+391	+8*
$2\omega + p - \varsigma$	+60	+365	+8
$p - \varsigma$	+2303	+361	+7
$2\omega - p - 2\varsigma$	+336	+336	+7*
$p + \varsigma$	+231	−335	−7
$2\omega + p + \varsigma$	+97	−329	−7
$2\bar{\delta}$	−19618	+314	+6
$4\omega - 2p$	+1581	−276	−6
$2\bar{p}$	+37408	−261	−5
$2\omega - p$	−232557	−259	−5
$4\bar{\omega}$	+488	+242	+5
$2\omega - p + \varsigma$	+6118	−223	−5
$2p - 2\delta$	+2014	+219	+5
$2\omega - p - \varsigma$	+2309	+196	+4
$2p + 2\delta$	+194	+194	+4*
$4\omega - p - \varsigma$	−156	−156	−3*
$\bar{p} - 2\bar{\delta}$	−3489	+136	+3
$4\omega - 2p + \varsigma$	−120	−120	−2*
$4\omega - 2p - \varsigma$	+120	+120	+2*
$3\omega - p$	−97	−97	−2*
$2\omega - 4p$	−93	−93	−2*
$3p - 2\delta$	+90	+90	+2*
$4\omega - \varsigma$	+86	+86	+2*
$\omega - p$	+437	−82	−2
$2\omega - \varsigma$	−3615	−81	−2
ς	+33969	+67	+1
$4\omega - p + \varsigma$	+65	+65	+1*
$4\omega + p$	+182	+60	+1

Argument	Coefficient	Diff.	″
$2\omega + 3p$	+53	+53	+1*
$2\omega - 3p$	+1041	−52	−1
$2\bar{\omega}$	+103900	+50	+1
$2\omega + \varsigma$	−2787	+44	+1
$2\omega - 2\varsigma$	−64	−38	−1
$4\omega - 4p$	−160	−37	−1
$\bar{\omega}$	−7727	−36	−1
$2\omega - 2\delta$	+4199	+35	+1
$\varsigma - 2\delta$	+62	+32	+1
$4\omega - p$	+411	−30	−1
$2\omega + 2\delta$	−128	−26	−1
$2\omega + p - 2\delta$	+101	−23	0
$2\omega + p$	+2647	−18	0
$2\omega - \varsigma - 2\delta$	−481	−18	0
$2\omega - p - 2\delta$	−60	+16	0
$2\omega - 2p - 2\delta$	−87	−14	0
$2\omega - 2p$	−3024	−10	0
$3\bar{p}$	−1732	−10	0
$\omega + p$	+116	−10	0
$\varsigma + 2\delta$	0	−9	0
$p + 2\delta$	0	−8	0
$\bar{p}$	−1096471	−6	0
$2\omega + 2p$	−1236	−4	0
$2\omega + \varsigma - 2\delta$	+623	+4	0
$2\omega - 2p + 2\delta$	0	−3	0
$\omega - \varsigma$	+151	−2	0
$4\omega - 3p$	+586	+1	0
$\omega + \varsigma$	−1024	0	0
$3\bar{\omega}$	+71	0	0
2ς	−497	0	0

7.6 Conclusions

Ideally, the transformation presented in the last section would reproduce wantij exactly. As Display 7.2 shows, we seem to be on the right track, but there remain some notable differences between our candidate and Mayer's stated result. The differences are not entirely trivial and need explanation before we are allowed to conclude whether we have indeed reconstructed the procedure that Mayer followed.

First, let us make some remarks on precision. Our candidate reconstruction of Mayer's procedure reproduces Mayer's multistepped equAtions in *Theoria Lunae* to four or five decimal places. The differences are in all cases $10''$ or less, and most of them are even smaller than $5''$. The precision to which our results match Mayer's compares very favourably to the precision of his lunar tables, in which he was prepared to neglect equAtions smaller than $30''$. However, Mayer stated his results in *Theoria Lunae* in radians to seven decimal places, that is $\frac{1}{50}''$. Such a precision is far beyond the accuracy that could be expected of his lunar theory. Mayer had no statistical theory of errors, and the modern scientific consensus that the number of significant digits in a quantity should reflect its accuracy was not established in his time. We will return to this point of overprecision on p. 180 below.

However, if our reconstruction of Mayer's procedure were exact, then we should have obtained identical results. Discrepancies may arise from several causes, such as a slightly differing composition of the terms that make up $\mathcal{M}$, a different way of substituting the true solar longitude for the mean solar longitude, or a different annual equAtion of the node. Further, numerical details of Mayer's computations are likely to differ somewhat from *Mathematica*. He is likely to have made some errors in his calculations, his cut-off and rounding may be less consistent than in a computer program, and he may have left out many small terms of the computation deliberately, based on his numerical insight, whereas a computer follows rules strictly.

But then a more surprising reason emerges to explain why our results differ from Mayer's. Apparently our transform has yielded several terms, indicated by the asterisks in Display 7.2, that were absent from Mayer's list. These terms are produced by the approximation technique and they do not correspond with terms in the original single-stepped list. Looking again at Display 7.1, we notice that old arguments and new arguments stand in a one-to-one correspondence. Thus, we see that in Mayer's lists, combinations of arguments absent from the single stepped solution were also absent from the multistepped solution and conversely. We may conclude that the form of the arguments, rather than the magnitude of the coefficients, was Mayer's criterion for inclusion of terms in the transformed equation. Among the ones that he skipped are the four terms with the largest differences between the reconstruction and Mayer's original.

The differences in Display 7.2 show also some interesting symmetries. Take, for example, arguments $p + \varsigma$ and $p - \varsigma$: differences for their coefficients are equally large except for a reversion of sign. The same symmetry is found in several other cases, e.g., with arguments $2\omega \pm p \pm \varsigma$. At least some of them can be traced back to where we substituted the true sun for the mean sun, indicating that perhaps we should really have redone the whole theory with the true sun, instead of just applying the solar equAtion of centre (see p. 131). Although the true sun differs but marginally from the once-equAted sun, a derivation of the theory from scratch could perhaps have markedly different coefficient values due to round-off and truncation errors.

Remarkably, the overall magnitude of the differences is rather robust. Changing several minor details in the computer transform usually has hardly any effect on the root mean squared error of the differences. Such changes included for example, neglecting intermediate results smaller than a certain threshold (say, $30''$), changes in the composition of the minor equAtions, computation of second-order terms throughout, and amendments to parameters such as the eccentricity of the terrestrial orbit. But structural changes to the multistepped procedure, i.e., different interpretations of Mayer's text, make the results usually significantly worse than in Display 7.2. In fact, my understanding of Mayer's text improved considerably during these numerical experiments, until I arrived at the interpretation of this chapter, which is fully in accord with the steps of computation in the rede tables. The following three examples will serve as illustrations.

As a first example, initially some of my computed coefficients differed markedly from Mayer's values, even up to $20''$. The differences could be explained away by assuming just two misprints in the single-step solution of *Theoria Lunae*. Later I realized that the deviations were the result of my negligence of the sizeable second-order terms (7.12) and (7.13). Inclusion of the second-order terms provides a much more satisfactory solution than the supposed printing errors. This also suggests that my approach is generally sound.

Second, we have seen that the multistepped scheme equAtes the lunar anomaly, elongation, and nodal distance by application of the minor equAtions, $\mathcal{M}$. But with which coefficients do we take these minor equAtions? Should they be the coefficients that can be read off from the original untransformed single-step solution, or should first the true sun be substituted in the arguments? The *Theoria Lunae* text quoted on page 122 above suggested the existence of an intermediate equation of the lunar longitude, differing from the single-step one only by the true sun being used instead of the mean sun. If this were the case, then the $\mathcal{M}$-coefficients would naturally have been selected from the intermediate solution and not from the single-step solution. So I tried both alternatives in the computer program, and found notably smaller deviations with the coefficients of the minors from the intermediate solution, than from the single-step solution. This is a strong argument for the trustworthiness of Mayer's account in *Theoria Lunae*.

Then we get to the third example, concerning the place of evection. As was shown before, the kil tables of 1753 and the final rede tables give a different place to the evection. Because *Theoria Lunae*'s notation distinguished between ω and $\tilde{\omega}$, we may infer that by the time of its completion the evection had already been put in its definitive place, as in the final tables. Experiments with both places of evection in the computer transform showed decidedly that the multistepped scheme of *Theoria Lunae* had evection indeed in its final place. This agrees well with the date of November 1754 when evection was moved into its new position, and it also puts it beyond doubt that this part of *Theoria Lunae* did not exist before that date.

All things now taken into account, we are justified to conclude that Mayer indeed transformed the single-step solution into the multistep one, just as he claimed. Mayer's rather vague description in *Theoria Lunae*, combined with knowledge of his lunar tables, provides enough information to reconstruct a historically plausible procedure by which he transformed the former solution (*Theoria Lunae* §48, derived from the differential equations) into the latter (*Theoria Lunae* §49, reflecting his lunar tables). Our reconstructed procedure is close to the one that Mayer followed, except that perhaps he had introduced the true sun not by substitution, but by derivation of the theory from scratch. The approximation technique described in Sect. 7.4 must have been his tool to work the equation into its final form.

Chapter 8
'Hausbackene Combinationen'

The success of Mayer's lunar tables depended to a large extent on his artful adjustment of their underlying coefficients to observations. In his own view, expressed in the introduction of *Theoria Lunae*, the theory was incapable of producing all coefficients accurately, hence fitting was a necessity. The size of such a task must not be underestimated: Mayer fitted all 14 longitude equations plus a few provisional ones plus the epochs and mean motions to over a hundred lunar observations. With least-squares algorithms now omnipresent on today's computers, such a fit costs hardly more time than what is needed for the data entry. But the panorama looked completely different in the 1750s. Computers were workers of flesh and blood operating with pencil and paper, the method of least squares had not yet been invented, and even statistical reasoning about model fitting was in its infancy. One of the pioneers in this area was Tobias Mayer. He held a particular appetite for large data sets and their application to the modelling of reality: be it with regard to the position of the moon, a geographical map of Germany, or temperature distribution on the earth.

Saving some of those other activities with large data sets to the next chapter, our current interest will be directed to his work of fitting the lunar table coefficients to observations. What will emerge is that Mayer did not have a systematic method (such as least squares), although he had developed some tools which are highly interesting in themselves.

The organisation of this chapter is as follows. Section 8.1 shows that not much can be gleaned from existing printed sources of what exactly Mayer did in order to adjust the coefficients in his lunar theory to the observations. To learn more, we must turn to certain unpublished manuscript sources that are described in Sect. 8.2. Section 8.3 explains my methods to investigate these manuscripts, and concludes with examples of obtained results. In Sect. 8.4, I propose a general procedure by which Mayer worked, and which explains the intimate connections shown to exist between several of those manuscripts.

S.A. Wepster, *Between Theory and Observations*, Sources and Studies in the
History of Mathematics and Physical Sciences, DOI 10.1007/978-1-4419-1314-2_8,
© Springer Science+Business Media, LLC 2010

Section 8.5 is devoted to the analysis of the few manuscript pages where most of Mayer's work crystallized. These pages are akin to modern spreadsheets. The results are presented and reflected upon in Sect. 8.6.

8.1 A Search in the Literature

In order to try to get a grip on the fitting of coefficients to tables, we start our information gathering enterprise with a search in the existing literature for clues: first in Mayer's published work, then in commentaries.

8.1.1 Mayer About His Own Work

The logical place to start is the most prominent of Mayer's writings to the most prominent astronomer to judge his work: *Theoria Lunae*. In its introduction, the author acknowledges *that* he corrected (in his own terminology) the coefficients to observations, but not how. Later in his text, he makes a more elaborate statement, devoting a complete paragraph entitled '*Correction of this formula through observations*' to it:

> After I had compared this formula [wantij, cf. Display 7.1] with many observations, it easily appeared that the eccentricity of the lunar orbit, which I had supposed in this calculation = 0.05454 (§40), should be increased with its 1/300th part, and indeed that instead of the first term of this formula $-6°16'56'' \sin \tilde{p}$ should be substituted this: $-6°18'11'' \sin \tilde{p}$. Further also the term $-2'38.6'' \sin \tilde{\omega}$, which is the largest of those that depend on the solar parallax, was found to be too large, since the observations do not allow it to be greater than $-1'55'' \sin \tilde{\omega}$. Moreover, many terms appear in this same formula which the theory, even if treated with the utmost care, cannot furnish accurately, for reasons well known to anyone who has exercised his vigour and patience in this matter. Therefore I have preferred to define these ambiguous terms from the observations, rather than for their sake to pursue the extremely tedious calculations still further (which I have been undertaking for a long time, with greater accuracy than anyone before me so far), and to augment them by new and perhaps insurmountable difficulties. Finally I have also corrected, through observations, those terms that the theory reveals with sufficient accuracy, to have them agree better with the heavens, given the addition or subtraction of a few seconds. I point this out principally for this reason, that nobody should think that the calculation set forth so far, or the formula derived from this in the preceding §, were more accurate and more in conformity with the observations, than my corrected lunar tables. For just as that calculation rests only[1] on the theory and is pressed upon from all sides by many difficulties, so are the tables built upon both theory and practice, and so they must necessarily be preferred. Meanwhile however, in order that it be clear how little the tables differ in most inequalities from this calculation, let me add this comparison of them. [follows list of wad (unfitted) and

---

[1] The printed edition has *Soli*, which makes no sense, and is here interpreted as *soli*. In the manuscript, *Soli* and *soli* are indistinguishable, and Mayer was not strict in capitalizing names of celestial bodies.

wijd (fitted) equAtions; see appendix A] The terms that the formula printed in §49 [wantij] embraces besides those [that are printed in the list], are partly in themselves very small by themselves, and partly in fact made dubious by the theory. But after a most diligent examination of observations had been undertaken, it appeared sufficiently evident that they are to be plainly rejected from the tables.[2]

So here we learn not only that Mayer fitted the coefficients of his lunar equAtions to the observations, but also by how much. He expressed that it is more desirable to have tables based on both theory and practice, than on theory alone; partly because some of the coefficients cannot be accurately computed from the theory (this is due to small divisors in the fractions that determine them; see the text leading up to fn. 57 in Sect. 5.4.10). He rejected the smallest equAtions, mainly because comparison to observations had shown him that the tables could do without them. But Mayer does not tell his reader how he managed to adjust his coefficients to observations.

Surprisingly, this quotation exhausts all the evidence about fitting of his lunar equAtions that Mayer ever had intended to go into print. Perhaps then we find some information in his correspondence, first of all with Euler.[3] From these letters emerges that Mayer had some kind of a lunar theory based on gravitation at least since January 1752 (although his later adoption of the *NTM*-like structure and his troubles to complete *Theoria Lunae* cast doubts about its merit). Already at that time he used observations to fit some of the coefficients of the equAtions. More telling of Mayer's methods is a letter

---

[2] '*Correctio huius formulae per observationes.*

§50 Postquam hanc formulam cum observationibus pluribus contulissem, facile adparuit, eccentricitatem orbitae lunaris, quam = 0, 05454 in hoc calculo supposueram (§40.) augendam esse parte sua 1/300, atque adeo loco termini huius formulae primi $-6°.16'.56''$ *sin* $.\tilde{p}$ *hunc esse substituendum* $-6°.18'.11''$ *sin* $.\tilde{p}$. *Deinde etiam terminus* $-2'.38''$, 6 *sin* $.\tilde{\omega}$, *qui maximus est eorum qui a parallaxi Solis pendent, nimis magnus est deprehendus, quippe quem observationes maiorem non admittunt, quam* $-1'.55''$ *sin* $.\tilde{\omega}$. *Praeterea in eadem hac formula plures termini occurrunt, quos theoria, licet summo studio tractata, accurate praebere non potest, ob rationes nulli non cognitas, qui in hac re vires ac patientiam exercuit. Hos igitur terminos dubios malui ex observationibus definire, quam illorum gratia calculum taediosissimum, a me iamdudum longe accuratius institutum, quam a quoque hactenus factum, ulterius adhuc persequi, novisque ac forte insuperabilibus augere difficultatibus. Denique etiam eos terminos, quos theoria satis manifeste ostendit, per observationes ita correxi, ut paucis secundis adiectis vel demtis cum coelo magis consentirent. Haec ideo praecipue moneo, ne quis putet, calculum hactenus expositum seu formulam § praeced. inde derivatam accuratiorem esse magisve conformem observationibus, quam tabulas meas lunares correctas. Ut enim iste calculus soli* [see previous fn.] *theoriae innititur, plurimisque undique premitur difficulatibus; ita tabulae utrique tam theoriae quam observationibus inaedifiuutae sunt, atque adeo non possunt non esse praeferendae. Interim tamen ut pateat, quam parum tabulae a calculo hoc in perisque inaequalitatibus differant, subiungam hanc earum comparationem.* [follows display of coefficients] *Termini, quos praeter hos formula §49 exhibita complectitur, partim in se minimi sunt, partim vero a theoria dubii relinquuntur. Observationum autem diligentissimo examine instituto satis evidenter patuit, eosdem ex tabulis prorsus esse reiiciendos'* (Mayer, 1767, pp. 50–53).

[3] Forbes (1971a).

of him to Euler, written on 6 March 1754. In it, Mayer asserted that he
had managed to correct the accuracy of his tables' predictions to within 30″,
which was twice as accurate as lunar positions obtained through meridian
transits. He was only able to achieve this remarkable feat by doing away
with that kind of observations. He had turned to the much more accurately
observable occultations of stars by the moon and explained:

> In order, however, that the uncertainty in the place of the fixed star might have no
> appreciable influence on the position of the Moon, I have first of all collected only
> the occultations of a single fixed star; then also if there were already a few minutes
> of doubt in the position of this fixed star, the error would certainly not prejudice
> the use of these observations in the investigation of the inequalities of the Moon's
> motion, but merely be reflected in the epoch of the mean motion of the Moon, which
> one can afterwards easily correct through the observations of a solar eclipse together
> with the position of the fixed star.[4]

He said that he had first made use of occultations of Aldebaran of which a
considerable number were available to him. Thus having found the most significant corrections of his tables, he similarly made use of the occultations of
other bright stars such as Spica, Regulus, Antares, and the Pleiades. All in all,
he compared his lunar motion model to over a hundred observed occultations
and obtained

> ...still more, although almost inappreciably small, corrections of those equations
> which I employ in my printed tables.[5]

With the tables thus corrected, Mayer was able to detect errors of sometimes
several minutes in the 'general', i.e., meridian, observations of the moon.

Illuminating as these words may be with respect to Mayer's penetrating
expertise and dedication to create the best possible data sets, they are reticent
regarding the inference of the best fitting parameters from those data. Further
down (Sect. 8.6) we will have the opportunity to connect Mayer's report
with his computations. Other published correspondence of Mayer, however,
is silent about the fitting of coefficients, and the same holds for the several
volumes of posthumously published manuscripts.[6]

8.1.2 Commentators of Mayer's Work

We now turn to various commentators on Mayer, and first of all to Jean le
Rond d'Alembert. The first volume of his *Recherches sur differens points importants du système du monde*, appearing in 1754, contained a sceptical note
on Mayer's kil tables. He doubted their theoretical foundation and distrusted
the accuracy claimed for them, primarily because Mayer's tables held so small

---

[4] Forbes (1971a, p. 81).

[5] Forbes (1971a, p. 81).

[6] Forbes (1971c, 1972, 1983) and Forbes and Gapaillard (1996).

a number of equAtions compared to, e.g., Clairaut's. He challenged Mayer to supply observations in support of his claimed accuracy, and was promptly served by a list of 139 lunar positions of Bradley compared to Mayer's tables, which Mayer published in the third volume of the Göttingen *Commentarii*.[7]

d'Alembert was not quite convinced. Although he could not escape the conclusion that Mayer's tables seemed more accurate than those of Lemonnier and Clairaut, he insisted that a larger volume of data was needed before any definitive conclusions could be drawn. After all, it had taken many years before it became apparent that Newton's *NTM* was often less accurate than the 2′ claimed for it. d'Alembert adhered to the viewpoint that Lemonnier's lunar tables (incarnating the *NTM* prescriptions) were to be preferred to Mayer's because they had been compared to observations over a longer time, and consequently their error behaviour was better known, than Mayer's – a healthy viewpoint indeed.[8] As we have seen, d'Alembert's doubts about the theoretical basis of the tables were justified, but Lemonnier's tables were ultimately abandoned in favour of Mayer's.

D'Alembert's initial scepticism of Mayer's pragmatic approach was soon moderated. Even d'Alembert had to admit that theory left some equAtions insufficiently well determined and that observations were necessary to fill in the blanks:

> However, the uncertainty of the coefficients that the Theory provides, compels to have recourse to observations.[9]

He also held ideas how that should be done: he proposed to select observations according to carefully prescribed configurations of the sun, moon, and earth as well as the apsides of their orbits, in such a way that the resulting system of 19 equations breaks down into easily manageable portions before the 19 coefficients are determined. He seems not to have practised his proposal with real observations.[10] Moreover, his method is deterministic, not statistical like Mayer's is, as we will see later. D'Alembert uses just as many observations as are necessary to fix every parameter. Mayer, on the other hand, was searching for ways to include more observational evidence in the determination of his parameters. He subscribes more readily than d'Alembert to the basic facts that there are chance errors in the observations, that these errors are bound to influence the parameters, and that they have a certain tendency to cancel each other.

A few decades later, Jean-Baptiste Joseph Delambre expresses himself considerably more enthusiastic about Mayer's work. Delambre justifies Mayer's

---

[7] d'Alembert (1754–1756, I, p. xxvi). Mayer's comparisons to observations were appended to Mayer (1754). Additional comparisons were later printed in Mayer (1770, pp. 30–31).

[8] d'Alembert (1754–1756, III, pp. 43–45). Bradley, by the way, was to undertake a comparison of Mayer's *improved* manuscript lunar tables with a significantly larger volume of observations.

[9] *«Or l'incertitude des coefficiens que la Théorie donne, oblige de recourir pour cela aux observations»* (d'Alembert 1754–1756, III, p. 44).

[10] d'Alembert (1754–1756, III, pp. 53–63).

use of observations: the coefficients of some equAtions, he says, are obtained from theory with insufficient accuracy, and observations are needed to determine them properly. Observations are required also to fill in the fundamental constants such as the masses of the bodies and the epochs and eccentricities of unperturbed orbits. These constants cannot be known merely through the solution of the differential equations, but, to phrase it in modern terms, they are fixed by the boundary conditions of the system of differential equations. Delambre emphasizes to make the best possible use of the observations and is willing to fit the magnitudes of all the perturbations directly from them, instead of indirectly via the integration constants in the theory.[11]

Delambre also asserted that

> Mayer adopted in [his] theory only the form of the arguments, and determined the coefficients by the observations. He adhered mainly to solar eclipses and occultations of stars, which require no other instruments but a telescope and a well regulated clock; moreover there was no collection of [observations of] meridian passages.[12]

Delambre's last remark is curious, because there *were* such collections made at the Greenwich Observatory, and Mayer had access to part of them. Flamsteed's observations had been published,[13] and Mayer had received a transcript of 139 of Bradley's observations via Euler. There were also series of meridian passages observed in Paris. One can hardly imagine that Delambre was unaware of the existence of all of these observations.[14]

Neither d'Alembert nor Delambre were specific about Mayer's fitting, and they could not have been more specific given the lack of information from Mayer himself. More is to be expected from those who have had access to the manuscripts left behind by Mayer, primarily from Eric Gray Forbes. Regarding Mayer's statement in *Theoria Lunae* that he had adjusted the theoretical coefficients from observations, Forbes and his co-author Curtis Wilson remarked:

> For so massive an operation of correction, it is plausible to assume that Mayer made use of equations of condition – the only then known statistical procedure for

---

[11] Delambre (1827, p. 445).

[12] «*Mayer ne prenait dans la théorie que la forme de ses argumens, et déterminait les coefficiens par les observations. Il s'attachait principalement aux éclipses de Soleil et aux occultations d'étoiles, qui n'exigent d'autres instrumens qu'une lunette et une horloge bien réglée; d'ailleurs on n'avait aucune collection de passages au méridien*» (Delambre 1827, p. 440).

[13] Flamsteed (1725).

[14] There is ample evidence that Mayer used the Paris observations. He copied 13 meridian transits from the *Mémoires* for 1739; his extract of that volume is contained in Cod. μ_{12}. He used these observations in Cod. $\mu_{41}^{\sharp}$ to demonstrate the quality of his printed kil tables, and listed the results (i.e., the differences between computed and observed positions of the moon) in the introduction of the kil tables. Delambre might have been unaware that Mayer had used these data, but he is unlikely to have missed Mayer's published comparison of his tables against Bradley's observations appended to Mayer (1754).

determining a large number of unknowns simultaneously, and one which Mayer alone had previously used with success.[15]

What was there called 'equations of condition' is also known as the method of averages. Though not of his own invention, Mayer had applied it on an earlier occasion, in an effort to determine the orientation of the lunar rotational axis. In short, the procedure provides a way to handle an overdetermined system of equations relating a number of observations to a smaller number of unknown parameters. Mayer's successful application of the procedure, which he clearly exposited in the *Kosmographische Nachrichten*,[16] contributed much to the popularity of the method of averages in the second half of the eighteenth century. The procedure will be discussed more fully in Sect. 9.4.2. Forbes and Wilson added that

No doubt the accuracy of Mayer's tables *was* primarily due to the skilful fitting of theory to observational data, but Mayer's artfulness in such empirical determinations of numerical coefficients was new and unmatched among his contemporaries.[17]

This artfulness was certainly new and unmatched, as was, by the way, the sheer accomplishment of such a massive fit. It is no unreasonable hypothesis that Mayer applied the method of averages again under these circumstances. But nobody *knows* how, in any more detail, Mayer went about to adjust his coefficients. As I will show, he did not apply his method of averages in this case: the operation is perhaps too massive for it to be used expeditiously, and there are too many variables.

To conclude this brief survey, we return to Göttingen, where early in the nineteenth century a new observatory had been built. Its director and inhabitant was Carl Friedrich Gauss, who was literally living on top of the manuscripts of his predecessor Tobias Mayer, and from whom we may therefore expect a well-informed opinion. Gauss, characteristically, was not so much impressed by the massiveness of Mayer's fit as he was disappointed by the lack of a method. Reflecting on the invention of the method of least squares (of which he claimed priority over Legendre), he remarked to his friend Heinrich Christian Schumacher:

...I do not doubt that already many of the mathematicians in that category, e.g., Tobias Mayer, have cared to help themselves with that practice [i.e., least squares] in circumstances where it would have been worthwhile. I would have been ready to make a bet for this opinion. – Though now I know – that I would have lost this bet, because since a few years papers of Tobias Mayer [are] in possession of the observatory, [which show] that Tobias Mayer computed *not* according to a systematic principle, but only according to homely combinations.[18]

---

[15] Forbes and Wilson (1995, p. 66); Gautier even thought it was so plausible that he presented it as a fact in Gautier (1817, p. 73).

[16] Mayer (1750a).

[17] Forbes and Wilson (1995, p. 66).

[18] *"[...] ich zweifle nicht, dass schon längst viele unter jener Cathegorie gehörende Mathematiker, z.B. Tobias Mayer sich in Fällen, wo es der Mühe werth gewesen, jenes*

It is time now that we turn to those papers of Mayer's to see how much of
Gauss' statement is true.

8.2 A Variety of Manuscripts

The main results in this chapter are derived from a complex set of interrelated
manuscripts, which deserve a detailed introduction before their relations will
be discussed. The relations are by no means immediately apparent, yet they
are essential to unravel the process of fitting that Mayer employed. So first
now, the relevant manuscripts will be presented; more manuscripts are listed
in appendix C.

Cod. $\mu_{41}^{\sharp}$: *Etwa 300 Bl. über Sternbedeckungen und Mondbeobachtungen,
frühere und eigene Beobachtungen und Rechnungen.* The bulk of this manuscript of 742 quarto pages consists of computations of lunar positions from
lunar tables, for specified times, and compared to observed positions. The
manuscript contains about 366 such computations based on either lunar
tables or coefficient sets (it is not always evident whether Mayer performed
his computation with the help of tables, therefore I will not always make
a clear distinction between tables and coefficient sets, often naming only
one and implying the other too). An example of such a calculation (*position
computation* for short) is in Fig. 8.1. Each of the computations started with
a date and time for which an observation was available in Mayer's records.
He computed the position of the moon as predicted by tables (or a set of
coefficients), compared the result to the observed position, and inferred an
error of the tables, i.e., a difference between calculated and observed place
of the moon.

The calculations cover a very interesting period in Mayer's efforts around
the publication of the kil tables, and they are extremely relevant to our understanding of his research at that time. They range from engagement with
the follow-ups of the 1752 *Entwurf* (discussed above on page 112), to a computation to verify a reputed solar eclipse observable in China in 2128 BC, a
remarkable computation that he made following a request from Euler early

---

*Verfahrens bedient haben möchten. Ich würde für diese Meinung bereit gewesen sein eine
Wette einzugehen.—Allerdings weiss ich jetzt—dass ich diese Wette hätte verlieren müs-
sen, da seit einigen Jahren Rechnungspapiere von Tob. Mayer im Besitz der Sternwarte,
dass Tob. Mayer nicht nach einem systematischen Princip, sonder nur nach hausbacke-
nen Combinationen gerechnet hat",* Gauss to Schumacher, 24 June 1850 (Peters 1863,
p. 90). In a letter to Olbers dated 24 January 1812, he conjectures *"dass z. B. [Mayer] so
etwas angewandt hat, ohne es zu proclamieren",* ('that e.g., Mayer had applied something
like that [i.e., a least-squares method], without proclaiming it') which suggests that he had
not yet seen Mayer's manuscripts then. References to Gauss' correspondence came to my
attention via Sheynin (1993, p. 43) and Sheynin (1979, p. 23).

in 1754.[19] The calculation of the circumstances of this very ancient Chinese eclipse is furthermore remarkable because it is the only one in the whole manuscript that is based on the lunar tables contained in Cod. $\mu_{49}^{\sharp}$ to be discussed below. This period (1752–1754) covers the creation and fine-tuning of Mayer's kil tables, and consequently we find many of the coherent researches reported on in the introduction to those tables or in letters to Euler. There are also position calculations which form the starting material for part of Cod. $\mu_{33}^{\sharp}$, to be discussed next.

The folios of the manuscript are made up in quires, loosely stitched together to form a book-like whole. The ordering of the quires is not chronological. One of the quires makes the impression that it is particularly out-of-sequence, because it contains position calculations based on Euler's early lunar tables of 1746.[20] Some additional features of this quire make it rather interesting, but we postpone further discussion of these to Sect. 9.5.

Cod. $\mu_{33}^{\sharp}$: *Berechnung der Mondbeobachtungen, Vergleichung mit der Theorie.* This item comprises about 30 loose sheets, most of them of folio or double folio dimensions. Each sheet was made up in rows and columns. Every row answers to one date which we find written in the leftmost column, and further it is filled with numbers. Examples of such sheets are depicted in Figs. 8.3 and 8.4. Similar sheets exist amidst the papers on lunar theory in Cod. $\mu_{28}^{\sharp}$.

These sheets, which have never been mentioned before in the literature, actually embody the meat of Mayer's process to improve the coefficients of his lunar tables. Parts of the sheets, and sometimes by no means the least important parts, are very hard to read due to their poor state, and also because Mayer scribbled insertions and corrections on them, while he crossed out other information. Their form and function is very much akin to the modern objects that we commonly refer to with the word *spreadsheets*: Mayer tried the effects of various amendments to the coefficients of his lunar equAtions using these sheets, in order to adjust the equAtions to observations.

Cod. $\mu_{49}^{\sharp}$: *Sonnen- und Mondtafeln.* This is a collection of four quires, of which each comprises a separate set of lunar tables, similar in layout to the kil tables published in the *Commentarii.*[21] The first quire of the bundle (val) is drawn up in an orderly handwriting. The second quire (veen) is remarkable for two reasons: first because, contrary to the other quires in this manuscript, its evection table is the fifth, signalling a later stage of development (cf. remarks on the place of evection in Sect. 4.6) and second because half of its two-page evection table ended up with the Astronomer Royal in Greenwich.[22]

---

[19] For more details, see Forbes (1971a, pp. 79, 80, 84) and *Vorlesungen über Sternkunde* (Forbes 1972, I, pp. 96–97).

[20] Euler's tables were published in *Opuscula varii argumenti I* (Euler 1746). Mayer's own manuscript copy of these tables are preserved in Cod. $\mu_{14}^{\sharp}$, fol. 1–13.

[21] Mayer (1753b).

[22] It is now preserved in Cambridge University Library as a lone sheet in the Royal Greenwich Observatory Archives, Papers of Nevil Maskelyne, RGO 4/125. The dislocation

The third and fourth quire (vlakte and vlije, respectively) are in a less-orderly handwriting. These two sets of tables are each incomplete by themselves, but together they seem to form a complete set. Both are folded inside out, and one has been stuck in between the pages of the other, just as if they had been in frequent use for some time, complementing each other. I did not immediately recognize this condition since they had been lying in the archive, folded in that particular way, for much longer than their operational life in Mayer's hands, and the folds in the paper had naturally adjusted themselves to this permanent disordering. A shortlist on one of their pages summarizes the results that were obtained in a sheet of Cod. $\mu_{33}^{\sharp}$.

Cod. μ_5: *Berechnungen über den Mond zur Prüfung der Mayer'schen Mondtafeln.* This is a bound quarto volume completely filled with about 144 numbered position calculations, dating between 1661 and 2 October 1754. Numbers and dates correspond to numbers and dates on sheets 9–11 in Cod. $\mu_{33}^{\sharp}$. Analysis of a sample of the calculations showed that they are based on the mui and vlakte table versions, similar (but not identical) to the rak manuscript tables which Mayer sent to London in December 1754. The calculations also show that some improvements to the tables were made, which led to different, usually smaller, errors. As proof of the quality of the tables, a subset of the results obtained in this manuscript was apparently sent with the longitude packet to London, where they were eventually printed.[23] It is very likely, therefore, that these position calculations were made in the final stages before sending off the rak tables which formed the most significant part of his longitude method.

On one of the first pages, Mayer started a catalogue of stars (*loca fixarum supposita, initio anni 1750*) but he got no further than two: notably, Aldebaran and Regulus. These two stars played a major role in the improvement of coefficients (cf. page 146 and Sect. 8.5).

Cod. μ_7: *Berechnungen und Beobachtungen meist von Fixsternen.* This is a bound quarto volume, too. It contains various astronomical investigations, somewhat in the same vein as Cod. $\mu_{41}^{\sharp}$, though less voluminous, and it may be considered as a sequel to the latter. Its contents include a determination of the orbit parameters of Mercury from observations of its 1753 transit before the sun. There is a list of conclusions drawn from one of the spreadsheets, though apparently not one of the sheets in Cod. $\mu_{33}^{\sharp}$. These conclusions, as well as several lunar position calculations and a list of new lunar table coefficients, clearly link this manuscript to the era of the vlakte and mui table versions, just like manuscript Cod. μ_5 above.

---

of this page is explained by the circumstance that it held a revised table of secular acceleration on the backside, and Mayer had stipulated that the revised secular acceleration must be sent together with his 'last manuscript tables' to the Astronomer Royal after his death (also see RGO 4/130).

[23] Forty-one results (some with slightly differing errors) were included in Mayer (1770, pp. 30–31), along with Mayer's descriptions of the method of, and the instrument for, taking lunar distances.

8.3 A Pair of Pliers

Manuscript Cod. $\mu_{41}^{\sharp}$ counts approximately 366 position calculations. Some of the dates for which these were made recur multiple times, suggesting that Mayer performed the computation several times from different table versions. Some calculations are very brief or not even finished; others cover several pages, e.g., when they involve a widely observed eclipse offering abundant observational data to compare with the tabular predictions. Only about 45 relate to dates before the invention of the telescope in 1608.

Not only these computations, but also much of the material in between them put it beyond doubt that Mayer was testing and/or improving the accuracy of his lunar tables. Only rarely does he explicitly jot down with which version of the tables he is working. For example, at one point he declares that from thence he will use the kil printed tables;[24] while several calculations elsewhere in the manuscript are accompanied by a note that they were corrected using those printed tables. Most of the time there is no direct evidence at all.

So, in order to track the progress that Mayer made, it is necessary to have at our disposal a tool by which we can link position calculations to the tables (or coefficients) on which they were based. The remainder of this section will be devoted to a description of such a tool.

The tool consists of two parts: one part to extract the coefficients of the equAtions used in a position calculation and the other to extract coefficients from tabulated equAtions. I applied the second part to all extant versions of Mayer's lunar tables. The results, together with the coefficient sets that Mayer had enumerated in manuscripts or correspondence, resulted in an overview of all known coefficient versions (cf. Display A.1). The two parts of the tool together act as a pair of pliers: they show which tables were used when and where and they show which calculations were made with what table version. With these pliers I could, for example, confirm the truth in my suspicion that the quires of Cod. $\mu_{41}^{\sharp}$ were not bound in their chronological ordering.

8.3.1 Extraction of Coefficients from Position Calculations

Before proceeding to the tool description, it might be a good idea to reread the start of Sect. 4.4 and recall the overall procedure for the calculation of a lunar position. Throughout the various versions of lunar tables, there are some variations to the procedure (such as the placement of evection), but they need not distract us at this moment. Among other numbers, a position calculation shows (a) the true longitude and mean anomaly of the sun; (b) the

---

[24] 'In seqq calculi omnes subducuntur ex tabb novis lunae Solaribus impressis in Com.Soc reg gött. Tom.II.' Cod. $\mu_{41}^{\sharp}$ fol. 224v.

three lunar mean arguments of longitude, anomaly, and node; (c) the values
of the arguments of all applied equAtions; and (d) the values of the equAtions
themselves.

Figure 8.1 shows an example of Mayer's calculations in the manuscripts,
which we will use to illustrate the coefficient extracting part of our tool.

The example shows, near the top of the calculation, solar arguments
of $9^s12°25'32''$ and $6^s3°53'13''$, and the lunar mean arguments longitude
$2^s7°50'35''$, anomaly $0^s24°55'52''$, and node $5^s2°17'39''$. Mayer rounded these
values to arc-minutes before he used them as arguments to equAtions; his
rounding was not as systematic as is customary today.

On the left side of the calculation is a small subtable with the ten minor
equAtions, complete with argument values. The minor equAtions are numbered with Roman numerals, but Mayer changed the numbering of equAtions
several times. It is therefore not directly evident from either the Roman nu-
meral or the value of the argument which equAtion is intended in every case.
For example, it is a priori unclear whether the IIId equAtion with argument value $3^s17°0'$ in Fig. 8.1 refers to the equAtion with argument $2\omega + \varsigma$,
$2\omega - \varsigma$, or still a different one. But since all arguments are composed of simple integer linear combinations of mean positions, it is usually easy to detect
which combination yields the correct value of the argument (in our example:
$2\omega - \varsigma \approx 3^s17°0'$) and thus to identify the equAtion.[25] After all equAtions
have been identified in this way, we know the sequence of the equAtions.

Moreover, with each equAtion we know not only the value of its argument,
but also the value of the equAtion itself. Most equAtions are of the form
$e = c\sin\alpha$, where e denotes the equAtion's value, α its argument, and c its
(yet unknown) coefficient. For these equAtions we can, in principle, deduce
the coefficient value c from

$$c = \frac{e}{\sin\alpha}, \qquad (8.1)$$

provided that $\sin\alpha \neq 0$. Our example has a IIId equAtion whose value is
$-59''$; this would yield $c = -59''/\sin(3^s17°0') \approx -61.7''$. Now we must give
due consideration to the rounding which took place in both α and e. Mayer
rounded the tabular argument α usually to minutes of arc, while the table
provided the equAtion value e to arc-seconds. Except when $\sin\alpha$ is close to
zero, the rounding of the argument α to minutes of arc is insignificant, because
the coefficients are generally small. The rounding of e is more significant. In
place of computing a *value* for c, we do better to compute a *range* within
which we expect c to lie, given the values of α and e. Disregarding rounding
errors in α, we get a rough estimate of the table coefficient c from

---

[25] Complications arise from two causes: first, because an ambiguity may arise when several
linear combinations of the mean positions yield approximately the same value, and second,
because of computational errors or illogical rounding on Mayer's behalf. Both complications
can be resolved by looking at position calculations on neighbouring pages.

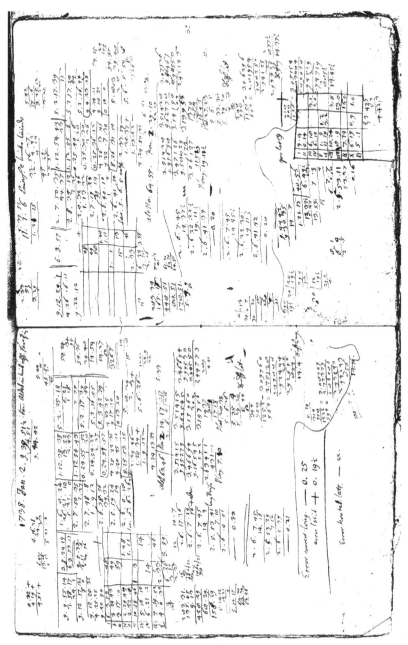

Fig. 8.1 An example position calculation, date 1738 January 2, $9^{\rm h}44^{\rm m}42^{\rm s}$ Paris mean time, inferred errors $-25''$ for longitude and $+19\frac{1}{2}''$ for latitude

$$c \in \left[\frac{e - 0.5''}{\sin \alpha}, \ \frac{e + 0.5''}{\sin \alpha} \right].$$

We can repeat this estimation for different α's from several different position calculations, and after a few repetitions the intersection of the intervals usually narrows enough to estimate the coefficient value c with an accuracy of $1''$ or better. We have to exercise care to select position calculations depending on the same coefficients, or else the intervals may have an empty intersection; but here the context of the calculations in the manuscript comes to help more often than not. For our example in Fig. 8.1 we get $c \in [-62.2'', -61.2'']$ using only one interval estimate, suggesting already that the IIId equAtion was $-1'2'' \sin 2\omega - \varsigma$.

Some equAtions, however, notably the equAtion of centre, the variation, evection, annual equAtion, and the equAtion depending on $\omega - p$, are more elaborate and answer to

$$e = \sum_{k=1}^{n} c_k \sin k\alpha \tag{8.2}$$

with, following Mayer's practice, $n = 3$ or 4 for the equAtion of centre and variation, and $n = 2$ for the other equAtions mentioned. For each of these equAtions, we can derive an estimate of Mayer's original coefficient values c_k by combining the data of at least n different position calculations and using a standard numerical solver. When selecting a set of calculations to determine the coefficients in these composed equAtions, care is needed again to ensure that Mayer performed them using the same tables. This is usually not difficult to establish; for example, many calculations were made batch-wise.[26]

In this way, we can gain rough estimates of all or most of the coefficient values in the lunar tables that Mayer was actually employing in a specific set of position calculations. Obviously, this coefficient determination breaks down in case $\sin \alpha$ is near zero, but usually it is possible to select another calculation from the same batch to make up for the near singularity. An evil exception is the variation coefficients, which can never be retrieved from eclipse computations.

8.3.2 Extraction of Coefficients from Tables

Now we turn to the other part of the analysis tool, the part that lets us extract the coefficient values from lunar tables. Let us assume that all equAtions are

---

[26] Real life tends to be more complicated. In a series of 18 position calculations clearly belonging to the same batch, one of the equations gave very doubtful and inconclusive results upon analysis. Suspecting an error in at least one of Mayer's computations, I repeated the determination 18 times, skipping each of the 18 (α, e) pairs from Mayer's computations in turn. The culprit soon gave itself away because in exactly one instance the results were satisfactory. An error was indeed detected in the skipped computation.

of the form of (8.2), with $n = 4$ and possibly with $c_k = 0$ for one or more values of $k \in \{1, \ldots, 4\}$. A tabulation of such an equAtion (as depicted in Fig. 4.1) is basically a list of pairs (α, e_α) consisting of arguments α and corresponding tabular values e_α, with α taking on values of whole degrees. We write e_α to explicitly express e's dependency on the argument. Given such a tabulation, we wish for a tool to retrieve the coefficient values c_k $(1 \leq k \leq 4)$. In the simplest form of an equAtion with only a single term, three of the c_k's will be zero.

From the table, we select five pairs (α, e_α) for five well-chosen argument values α. Four pairs suffice to extract the four sought-after coefficients; the redundant fifth provides a check. The values of α should be well spread over the range $0° \ldots 180°$. They should be 'easy' values (yet avoiding the zero points of the sine function) in the sense that $\sin(k\alpha)$ is easily obtained by geometry, so that we stand the best chance that Mayer computed the tabular value directly instead of via an interpolation procedure.[27] Still, the values e_α are rounded to seconds of arc in the tables, and this has a slight negative impact on the precision of the reconstructed coefficients c_k.

With appropriate values of α substituted in (8.2), it is easy to see that:

$$e_{30} = \tfrac{1}{2}c_1 + c_3 + \tfrac{1}{2}\sqrt{3}(c_2 + c_4), \qquad e_{150} = \tfrac{1}{2}c_1 + c_3 - \tfrac{1}{2}\sqrt{3}(c_2 + c_4),$$
$$e_{45} = \tfrac{1}{2}\sqrt{2}(c_1 + c_3) + c_2, \qquad e_{135} = \tfrac{1}{2}\sqrt{2}(c_1 + c_3) - c_2,$$
$$e_{90} = c_1 - c_3.$$

Of the various ways to solve these equations for c_k, we choose:

$$c_1 = \tfrac{1}{3}(e_{30} + 2e_{90} + e_{150}), \qquad c_2 = \tfrac{1}{2}(e_{45} - e_{135}),$$
$$c_3 = \tfrac{1}{3}(e_{30} - e_{90} + e_{150}), \qquad c_4 = \tfrac{1}{3}\sqrt{3}(e_{30} - e_{150}) - c_2;$$

different solutions might yield slightly different values for the c_k's. With these relations we can compute the four unknown coefficients from five tabular values. Again, we could do with one tabular value less, but the inclusion of a redundant tabular value provides an insurance, by way of back-substitution of the four coefficients c_k in the previous system, against rounding errors and blunders on behalf of either the creator of the tables or the person copying data from them.

This rough-and-ready procedure is a sensible alternative to a full-blown but time-consuming spectrum determination for each table, for instance by applying a Fourier technique capable of coefficient retrieval in much more complicated tabular structures. In those cases where the actual coefficients were known in an independent way, the procedure reproduced them within a margin of $1''$. Moreover, an incidental Fourier analysis of a sample of the

---

[27] Cf. Dalen (1993, p. 11). Van Dalen distinguishes between *nodes* which are precisely calculated table values and *internodal values* which have been interpolated.

tables yielded practically the same coefficients. A precision of $1''$ is sufficient to distinguish Mayer's different versions because his changes from version to version were larger than that.[28]

Applying this tool to every table of a lunar longitude equAtion discovered in the archive, I built up a collection of the various coefficients of the tables. To this collection I added the coefficient values found in the form of lists such as, e.g., those published in *Theoria Lunae* (cf. Display 7.1). This yielded a useful overview of the different versions of equAtions Mayer employed. The overview might still be incomplete either because I overlooked a still extant version, or because some versions might have been destroyed. Display A.1 lists all unearthed versions.

8.3.3 Examples

Now that we have provided ourselves with a means of extracting the coefficients that hide in a particular series of position calculations, as well as a complemetary means to recover the coefficients of the various versions of lunar tables, we are well able to match position calculations to table versions. Give or take one or two seconds of arc, a unique match can almost always be established. Now follow a few examples showing the power of this technique.

Our first example is taken from manuscript Cod. μ_1. On fol. 38v, there are three position calculations related to three solar eclipses. Recovery of the arguments of the minor equAtions shows that the evection is among them, so that they are of the later variety of tables. The Roman numbers of the equAtions match only with versions **rede**, **veen**, **wad**, and **wijd**. After extraction of intervals for the coefficient values, we must dismiss **wad** and **wijd** because almost all their coefficient values are slightly out of bounds. **Veen** matches (as far as we know its minor equAtions), except for one coefficient which is just slightly off. Two of the remaining four equAtions (equAtion of centre, variation, etc.) match **rede**, while **veen** is a few seconds off. A third equAtion is matched equally well by both. The fourth is variation which gives inconclusive

---

[28] D'Alembert used basically the same technique to retrieve the coefficients of the two separate terms that comprise Mayer's equAtion of the variation (d'Alembert 1754–1756, III, pp. 21–22). In our notation, d'Alembert assumed that Mayer's equation of centre equals $e_\omega = c_1 \sin \omega + c_2 \sin 2\omega$, whereupon he recovered the coefficients using $c_1 = e_{90}$ and $c_2 = e_{45} - \frac{1}{2}\sqrt{2}e_{90}$. The determination of coefficients that I deal with here is much simpler than in the cases presented in Dalen (1993). Van Dalen develops an array of tools to extract accurate coefficient values underlying astronomical tables with a more complex mathematical structure than in our case. Just because our tables represent sums of (at most four) sines of integer multiples of the argument, it is reasonable to expect that the tabular entries for arguments of $30°$, $45°$, etc., which we use to extract the coefficient values, are correct: computationally, they involve nothing more complicated than approximations to $\sqrt{2}$ and $\sqrt{3}$.

results: its argument ω is necessarily nearly zero at solar eclipses. We conclude that Mayer calculated these three position calculations with a table version at least nearly identical to rede.

Manuscript Cod. μ_5, our next example, contains about 131 numbered position calculations. Analysis of one of its calculations yields coefficient values that are met exclusively by mui, although what we know of mui is only a fragment of a complete set of tables and consists of a single sheet of paper with only six of the minor equAtions. The Roman ordinal numbers in the calcuIation match this mui sheet too. Identification by only six equAtions may seem doubtful, but there is spectacular additional evidence that Mayer indeed employed mui here. Mui's single sheet has two of the six tables on it corrected by a newer version which was glued over the original, and I analysed both old and new versions. Now, the magnitudes of the equAtions in the position calculation of Cod. μ_5 match the original version of mui, and the *difference* in magnitude between the old and new tables is written next to them! In other words: the position calculations agree with *both* the original and corrected versions of mui and they were altered according to the patched-on changes. This puts the use of mui beyond all doubt, even though we have only a limited number of coefficients to match. Further analysis of position calculations in the same volume shows that the six mui tables supplemented vlakte; indeed, mui and vlakte have an empty cross section of tabulated equAtions.

The third example deals with a long series of 55 sequentially numbered position calculations on fol. 192r–202r of manuscript Cod. $\mu_{41}^{\sharp}$. They have their ten minor equAtions sequenced as in the kil tables. Initially Mayer had computed the tenth equAtion of those position calculations (argument $\omega - p$) using zwin, but the results were crossed out and replaced by values based on kil. This puts the computations in the context of Mayer's transition from zwin to kil which took place some time between January and May of 1753 (zwin and kil differ mainly in the sequence of the equAtions, and in the coefficients of the equAtion depending on argument $\omega - p$). Here again we see an example of position calculations that were corrected following an amendment to the tables, just as in the previous example.

These position calculations are particularly interesting in connection with the Euler–Mayer correspondence and with Mayer's publications in volumes II and III of the Göttingen *Commentarii*. Mayer transmitted the coefficients of the zwin tables to Euler in a letter dated 7 January 1753. Kil represents the tables that were published in the spring of that year. When Mayer mentioned to Euler the publication of the latter, he added that he had made good use of some 139 of Bradley's lunar observations, which he had received a few months earlier. Euler lost no time to reply that he had personally sent those observations to Mayer.[29] The above-mentioned 55 calculations were based

---

[29] Euler had received these data from Gael Morris; he forwarded them via Christopher Schumacher to Mayer (Forbes 1971a, pp. 65, 68).

on these Bradley observations. In support of the accuracy of his printed tables, Mayer attached the results of these computations to an article on the applicability of his tables for longitude finding.[30]

As a final fourth example, analysis of the computations on folios 24r and 26r–29v in manuscript Cod. $\mu_{41}^\sharp$ reveals coefficients that somewhat resemble the versions kil, zwin, and gat, yet neither of these fits very well. Mayer performed these calculations using tables that I missed or that have been lost. I reconstructed all the equAtions and coefficients and dubbed them version put. After a more comprehensive study of the complete manuscript Cod. $\mu_{41}^\sharp$, I concluded that put and gat both represent phases later than the *Entwurf* of Display 6.2 but preceding zwin. Furthermore, some papers found in Cod. $\mu_{28}^\sharp$ document Mayer's experiments that led him from put to zwin.[31]

In these put computations, Mayer employed the same Bradley observations just mentioned, but the calculations showed considerably worse errors than the zwin and kil tables achieved. In view of Mayer's remark that he had received the Bradley data only a few months before May 1753, we get a ready impression of the speedy progress he was making in the preceding winter: he had conceived all versions gat, put, zwin, and kil during these months.

8.4 A Role for Each Manuscript

The examples just given in the preceding section, show how the pair of pliers helped to unveil the structure between Mayer's bequeathed manuscripts. As more and more connections between the manuscripts were found, a framework emerged in which these connections could be understood, modelling the process by which Mayer endeavoured to improve his lunar tables. Figure 8.2 illustrates this framework. Its constituents will be discussed in the present section. Some further details are filled in in Sect. 8.5. The key objects to keep track of are: theory, equAtions (with their arguments and coefficients), position calculations, and spreadsheets.

The theory box, top left, represents any form of lunar theory underlying the published (kil) tables and its derivatives. This theory was not the one in *Theoria Lunae*, but a mix of Eulerian techniques and elements from Newton–Lemonnier's *NTM*. Most of the material related to lunar theory is extant in Cod. $\mu_{28}^\sharp$.

The theory provided initial coefficients for the lunar tables or, alternatively, a sort of confirmation for coefficients derived in different ways (e.g., by trial and error). This is represented in the diagram by an arrow from the theory box to the coefficients and tables box.

---

[30] Mayer (1754).

[31] The papers are spreadsheets as discussed in Sect. 8.5, yet different specimens than those investigated there. All differences between the put and zwin versions were tried on these spreadsheets.

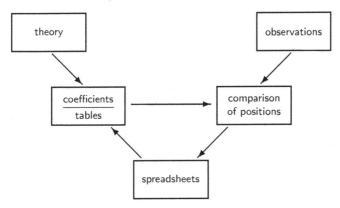

Fig. 8.2 Schematic presentation of the process of improvement of coefficients; the process (in particular the lower, triangular, part of it) is iterated several times

On the opposite side of the diagram we find the observations, which Mayer collected mainly in Cod. μ_{12} from varying sources such as *Philosophical Transactions of the Royal Society, Mémoires de l'Académie Royale des Sciences de Paris*, and private correspondence such as the list of Bradley's observations which he had received from Euler.

Mayer computed lunar positions from his tables and compared these with observed positions of the moon. This is indicated in the diagram by the two arrows ending at the comparison box. Many of these calculations were recorded sequentially in Cod. $\mu^{\sharp}_{41}$, Cod. μ_7, Cod. μ_5, and Cod. μ_1. Each comparison started with a date and time, and concluded with a table error, i.e., a difference between the calculated and the observed place of the moon. In this way Mayer was able to create a massive amount of data, from which he could judge the goodness of the tables. Moreover, he recognized that this aggregate of data would, in principle, enable him to improve the accuracy still further. But to actually carry out such an improvement must have appeared a rather daunting task to him, because he had 20–30 parameters to adjust, and there were random errors in the observations which could be, depending on circumstances, of the same order of magnitude as the errors in the predictions.

He found an ingenious way to handle this complex task. He collected and sorted the results of the position calculations, sometimes in between the calculations of Cod. $\mu^{\sharp}_{41}$, but more often on the separate spreadsheets now in Cod. $\mu^{\sharp}_{33}$ and Cod. $\mu^{\sharp}_{28}$. He then used these collections to try out various amendments to the table coefficients and he recorded the results. In this way Mayer was able to experiment and, glancing over the effects, to select the amendments that looked successful in diminishing the differences between predictions and observations. The essence of the process may be summarized in one sentence: Mayer fed the results of the comparisons into the spreadsheets, where he cooked up revised coefficients, with which he constructed new lunar tables. This is illustrated in the diagram by the arrows from the comparison box, via the spreadsheet box, back to the tables and coefficients box, completing a closed circuit.

We can identify several traversals of Mayer's through this circuit. In the first one, Mayer compared a subset of the list of Bradley's observations to position computations with coefficients of put.[32] He collected the results in spreadsheets,[33] where he experimented with changes to the put coefficients. This led to the zwin tables. After a few additional changes (not recorded on these spreadsheets), zwin was adjusted to form the printed kil tables.

In a later traversal, Mayer used the kil tables for the position computations, and he recorded the comparisons in Cod. $\mu_{41}^{\sharp}$. The first few spreadsheets of this round were also in Cod. $\mu_{41}^{\sharp}$; later ones were on separate sheets. Changes to the coefficients led to the generation of tables of the Cod. $\mu_{49}^{\sharp}$ family: val, vlakte, mui.

Mayer made another traversal using the new tables, recorded the comparisons in Cod. μ_5, and performed his experiments on spreadsheets 9, 10, and 11 of Cod. $\mu_{33}^{\sharp}$. Spreadsheets on folios 12 and up are of a later date. They must be associated with comparisons in Cod. μ_1 and parts of Cod. $\mu_{55}^{\sharp}$ in preparation of the 'last manuscript tables' rede.

Due to the volume and high information density of the archive material, I had occasion only to investigate one of his traversals in detail (the one that started from kil). Based on their visual appearance, it seems unlikely that the other manuscript folios reflect a significantly different procedure. On the other hand, a study of Cod. $\mu_{41}^{\sharp}$ is rewarding not only because of its link to the spreadsheets, but also because it extends from the preparation of the zwin and kil tables to their successors (the manuscript tables in Cod. $\mu_{49}^{\sharp}$), and, last but not least, because it contains many researches that Mayer referred to in texts and letters of that period, thus allowing us to date several of its parts.[34]

We will now turn to a more detailed analysis of the manuscripts themselves, in particular the spreadsheets.

8.5 Spreadsheets

Two examples of spreadsheets are depicted in Figs. 8.3 and 8.4. The first column of each spreadsheet lists calendar dates for which Mayer had made a comparison of observations and predicted positions (using tables of a kind). The second column contains the errors of the tables, i.e., the differences between the observed and calculated positions. The errors are obtained by

---

[32] These computations are in fol. 24–29, 74, 75 of Cod. $\mu_{41}^{\sharp}$.

[33] Double folio in Cod. $\mu_{28}^{\sharp}$, not paginated.

[34] Most notable are the investigations that he mentions in the foreword of Mayer (1753b). Several were also mentioned in his correspondence with Euler (Forbes 1971a).

subtracting the observed lunar positions from the calculated lunar positions. This, and the column headings such as '*Err. Tabb.*', indicate that Mayer regarded the tables, and not the observations, to be in error. That is a sound standpoint considering that his intention is to improve the predictive accuracy of his tables. In Sect. 9.1 we will investigate to what extent Mayer was aware of the errors present in the observations.

In order to get a first impression of the workings of the spreadsheets, we consider the columns following the initial errors. They usually contain a heading such as 'VI $= 1 - \frac{1}{4}$'. Such a heading signifies that the magnitude of the VIth equation was to be reduced by a quarter. The numbers in the column below the heading show what the *Err. Tabb.* column of errors would have been, if this revised VIth equation took the place of the original VIth equation.

The next column heading might be, for example, IX $+ \frac{1}{6}$ to indicate the increase of equation number IX by one-sixth. Usually the columns are cumulative, i.e., the errors tabulated in column n represent the errors as they would turn out after all the amendments indicated in columns $3, \ldots, n$. Thus, the numbers in the columns would at once show the differences between the observations (considered fixed) and the more or less plastic state of the tables, affected by amendments in this and all preceding columns. But sometimes Mayer decided to 'undo' part or all of the amendments made to the coefficients, and sometimes he revised a particular equation more than once on a sheet, so the cumulativity should always be checked carefully.

By way of example, the spreadsheet depicted in Fig. 8.3 will be discussed in detail in Sect. 8.5.1, but first we will take stock of the various sheets that play a role here.

Apparently, the first spreadsheets of the current round of improvements are those in Cod. $\mu_{41}^{\sharp}$, fol. 341v–342r, 43v–44r, and 46v–47r. Soon Mayer made them on separate sheets which are now located in Cod. $\mu_{33}^{\sharp}$ and Cod. $\mu_{28}^{\sharp}$. The numbers by which I refer to the sheets in Cod. $\mu_{33}^{\sharp}$ are those that have been assigned by library staff; these numbers do not necessarily respect the chronological order of the sheets. The sheets in Cod. $\mu_{28}^{\sharp}$ are not numbered; among those are two spreadsheets marked by Mayer as '*Tab. I*' and '*Tab. II*'. The latter is a continuation of the first, so they will here be referenced together as Tab.I/II. Only those spreadsheet folios will be studied here that play a role in the improvement of the printed kil tables towards rak, the first manuscript tables that Mayer sent off to London in December 1754.

Some other spreadsheets in Cod. $\mu_{28}^{\sharp}$ have already been mentioned on p. 160. They belong to Mayer's earlier improvement of put to zwin. Furthermore, I found that sheet numbers 9–11 of Cod. $\mu_{33}^{\sharp}$ were connected with position calculations in Cod. μ_5 which in turn were computed with the mui and vlakte tables, indicating that they date from the end of that period (see discussion of Cod. μ_5 in Sect. 8.2). The remaining sheets, numbers 12 and up, contain many data of the year 1757 and later, so they show that Mayer continued to use approximately the same technique to attain further improvements, most likely in connection with position calculations in Cod. μ_1 and the final rede tables. The technique clearly had a great appeal to Mayer.

This was work in progress to Mayer, and certainly he did not consider publication of the spreadsheets. Therefore, it is no wonder that he took no effort to explain or document the details of the enterprise in which they played a role. The notation 'VI $= 1 - \frac{1}{4}$' meant more to him than it means to us: we have to find out for ourselves the exact equAtion aimed at. We know neither its argument nor its coefficient, nay not even how many terms it comprises. EquAtions changed places, and coefficients varied, from version to version of the lunar tables. Although that complicates the task of analysis, it also provides valuable clues to cross-link the spreadsheets to tables and other manuscripts.

8.5.1 Example: Sheet 7

By way of example, we will now go through the columns of sheet number 7 of Cod. $\mu_{33}^{\sharp}$, reproduced in Fig. 8.3. The aim is to illustrate their content and to provide enough detail to allow an interested reader to check the computations for him- or herself. There is no intention here to convey a historically correct account of how Mayer worked with them: we will come to that point later in the chapter.

The dates in the left column of the sheet refer to 14 occultations (some perhaps appulses) of Aldebaran and 4 of Regulus. For these dates, the approximate mean solar anomaly ς and the ecliptic mean lunar longitude $\mathbb{D}$, elongation ω, and nodal distance δ are listed in Display 8.1, with an additional column 'err' that will be explained later. These values, rounded to degrees, come from modern computations using the kil epochs and mean motions, for the approximate time of day that Mayer stated along with his original position computation. They do not appear on the sheets but they are included here for ease of reference.

Ideally, the Aldebaran and Regulus occultations would all show a lunar longitude of $66° \pm 2°$ and $146° \pm 2°$, respectively, nearly identical to the longitudes of these stars but allowing some margin for the moon's semi-diameter, a time lapse of an hour or so, precession, etc. This precision is not always met, for reasons that I have not investigated further. It may also be noted that the dates of the Aldebaran occultations come in groups separated by about 19 years. This is because Aldebaran's ecliptic latitude is almost $-5\frac{1}{2}°$, which places it near the southern limit that the moon can ever reach in its orbit; consequently, Aldebaran occultations occur only under favourable positions of the ascending node, which cycles the ecliptic in 19 years.

The four columns following the date column have been crossed out by Mayer. Three of them are curiously numbered 11, 12, and 13; the first of these contains numbers whose source I was unable to trace. The headings

Display 8.1 Mean arguments of the data on Cod. $\mu_{33}^{\#}$, sheet 7, and the initial errors

Date	ς	$\mathbb{D}$	ω	p	δ	err
	°	°	°	°	°	$''$
Aldebaran						
1680 Sep 13	76	62	250	191	242	$15''$
1680 Nov 7	130	63	196	185	245	$93''$
1699 Aug 18	50	70	284	149	255	$60''$
1700 Jan 2	185	71	149	135	264	$49''$
1701 Feb 16	229	73	105	91	288	$-61''$
1717 Sep 25	86	70	247	132	245	$-18''$
1718 Feb 9	221	74	113	120	256	$-25''$
1719 Apr 22	292	72	40	70	278	$23''$
1719 Oct 30	120	69	212	46	285	$45''$
1737 Mar 8	248	73	85	64	265	$-24''$
1738 Jan 2	184	68	146	25	276	$-25''$
1738 Aug 8	39	65	289	358	284	$-15''$
1738 Oct 2	93	65	236	352	287	$98''$
1738 Dec 23	174	63	152	341	290	$12''$
Regulus						
1738 Dec 2	153	153	262	73	018	$25''$
1691 Feb 12	226	143	178	208	164	$1''$
1683 May 4	305	148	104	170	018	$32''$
1747 Mar 23	262	153	150	95	179	$8''$

and functions of each of the columns starting from the one numbered 12 are discussed below, using the top row for 1680 Sep. 13 as a running example and the equAtions of the kil version as reference. It is very enjoyable to compare the amendments that Mayer made in these columns to the differences between the kil and vlakte coefficients as listed in Appendix A; the reader may do this for him- or herself.

The heading of column 12 reads XII $= 1 + \frac{6}{60 \times 60}$. This means that the coefficient of the XIIth equAtion is increased by its 6/3600th part. The XIIth equAtion of kil is $-1°20'42'' \sin(2\omega - p) + 35'' \sin(4\omega - 2p)$, so that the amendment amounts to $-8'' \sin(2\omega - p)$, for all practical purposes. For the example date of 1680 Sep. 13, the amendment is $-8'' \sin(309°) \approx +6''$. Adding this to $-28''$ of the previous column gives $-22''$, which matches the entry on the manuscript (at least when we are willing to read it with its sign corrected).

Column 13 has a hard-to-read heading: either VI $= -\frac{1}{12}$, or perhaps $\frac{1}{10}$ or $\frac{1}{7}$. The VIth equAtion is $+90'' \sin(2\omega + p)$; $\frac{1}{12}$th of it is $7\frac{1}{2}'' \sin(2\omega + p)$. For the example date, we obtain $7\frac{1}{2}'' \sin(331°) \approx -4''$. Subtracting this from the value $-22''$ of the previous column yields $-18''$, as indeed we find on Mayer's spreadsheet. Diminishing by $\frac{1}{12}$ seems to be the correct interpretation.

The next heading reads V $= 1 - \frac{1}{4} - \frac{1}{20}$. The Vth equAtion is $+72'' \sin(2\omega - p - \varsigma)$; Mayer's notation here suggests that an earlier reduction of its coefficient by $\frac{1}{4}$ is now to be followed by an additional reduction of $\frac{1}{20}$ or 3.6''. Hence, the error value of $-18''$ from the previous column can now be updated and the new entry in the current column becomes $-18'' - 3.6'' \sin(2\omega - p - \varsigma) \approx -15''$.

Apparently, Mayer made a fresh start in the next column which he labelled III; it is not crossed out like the previous four columns, and it is separated from them by a vertical line. The additional note '*in libello occultat.*' ('in the little book of occultations') refers to the pamphlet that contains Mayer's position calculations (later included in Cod. $\mu_{41}^{\sharp}$). Our current column III also occurs on the spreadsheet of folio 43v–44r, where its values are computed by adding the same amendments as here to the initial errors of that sheet. For that reason, the initial errors of folio 43v–44r have been included in the Display 8.1 under the heading 'err'. To these, Mayer applied a constant shift of $-24''$ for the Aldebaran occultations, $-12''$ for the Regulus ones (thus effectively revising the star positions), and also an increase of the VIIth equAtion ($+58'' \sin(2\delta - p)$) by half of it, i.e., $29'' \sin(2\delta - p)$. Thus, the first entry of the column is $+15'' - 24'' + 29'' \sin(293°) \approx -36''$, as in Mayer's sheet.

The six columns IV–IX are similar and implement the following amendments, as can be checked by reference to the kil equAtions listed in Appendix A:

$$VI = 1 - \tfrac{3}{16}, \qquad \text{that is approximately } -17'' \sin(2\omega + p)$$
$$X = 1 + \tfrac{1.5}{60}, \qquad \text{that is } +10'' \sin(2\omega - 2p)$$
$$V = \tfrac{17}{20}, \qquad \text{that is } -11'' \sin(2\omega - p - \varsigma)$$
$$XII = 1 + \tfrac{6}{60 \cdot 60}, \qquad \text{that is } -8'' \sin(2\omega - p)$$
$$I = 1 - \tfrac{1}{120}, \qquad \text{or nearly } -6'' \sin(\varsigma)$$
$$IX = 1\tfrac{1}{6}, \qquad \text{or } -8'' \sin(2\delta - 2\omega)$$

The reader is encouraged to check that these amendments together, when applied to our standard example, yield the value $-7''$, whereas Mayer's result is $-8\tfrac{1}{2}''$. Such insignificant differences build up easily; they do not invalidate our basic understanding of what is going on.

The next two columns in Mayer's sheet implement a correction of the moon's mean motion, of roughly $+10''$ per 30 years after 1700. One column lists the correction and the next column (marked X) adds it to the accumulation in column IX. Column XI subsequently applies a shift of $-3''$ in the case of the Aldebaran occultations and $+2''$ to the Regulus occultations, which again reflect a revision of the ecliptic longitudes of these stars.

Variation is the object of the next seven columns. These columns list, successively, the two arguments ω and 2ω, then $-7'' \sin(4\omega)$, $-7'' \sin(2\omega)$, column $XI - 7'' \sin(4\omega)$, 3ω, and finally $+10'' \sin(3\omega)$. The argument $\omega = 250°$ for our example date is only $3°$ from Mayer's value $8^s 13°$, but the difference gets as large as $12°$ in the argument 4ω. It is easy to check the other values in the columns. Mayer finally chose only one amendment, $-7'' \sin(2\omega)$, and implemented it in the column marked as XII (the other columns, which he did not use to correct the error values, have not been included in Appendix B).

The next column looks as if it has been squeezed in. It lists $-7'' \sin(4\omega - 2\delta)$; these values are not used further. Then comes a list of $-7'' \sin(2p)$; neither have these values been used further.

EquAtion XIV, also known as the reduction to the ecliptic, is next. It is modified by $-\frac{1}{60}$, that is $-7''\sin(2\delta)$, which, again, is not used further on this sheet. The next column lists the value of argument $2\delta - p$, neither did Mayer use this here.

The final column is numbered XIII and it has an additional amendment to the VIth equAtion, on top of the amendment brought about in column IV. The equAtion is here further reduced by $\frac{1}{16}$, i.e., $6''$. In the case of our running example, this adds $+2''$ to the error of $-20\frac{1}{2}''$ in column XII, resulting in the final value of $-18\frac{1}{2}''$.

8.5.2 Data: Origin and Grouping

An inspection of the data and initial table errors that Mayer used in the various spreadsheets already enables us to put the sheets in a rough chronological order. Not much can be said about Mayer's criteria for admittance of specific observations into his data set, except that he preferred star occultations and solar eclipses observed with the aid of telescopes and pendulum clocks. There are only very few observations of Mayer himself, which comes as no surprise considering that the Göttingen observatory was not ready before the summer of 1754, and that Bird's mural quadrant was installed there only in February 1756.[35] We will now discuss the data on the spreadsheets.

Folios 341v–342r and sheet 8. The spreadsheet on Cod. $\mu_{41}^{\sharp}$, fol. 341v–342r, contains data from 15 solar eclipses. Almost the same dates in the same sequence occur on the 8th sheet in Cod. $\mu_{33}^{\sharp}$.[36] The initial error columns of the two spreadsheets match too. The remaining columns make it clear that sheet 8 is a continuation of fol. 341v–342r.

Folios 43v–44r, 46v–47r, sheet 6 and sheet 7. Cod. $\mu_{41}^{\sharp}$, fol. 43v–44r has 13 occultations of Aldebaran, a star then known as Palilicius. The same occultations, plus one extra, are re-used in spreadsheet Cod. $\mu_{33}^{\sharp}$ sheet 6. The initial table errors are the same in both cases, but sheet 6 also lists the values of several equAtions separately, which is rather uncommon.

These 14 Aldebaran occultations also appear in two other spreadsheets: in Cod. $\mu_{41}^{\sharp}$ on fol. 46v–47r and in Cod. $\mu_{33}^{\sharp}$ on sheet 7. The former has the same initial errors except in one place. The latter has no initial error column, but (as we saw in the example of Sect. 8.5.1) it continues from a particular column found somewhere in the middle of the former. Both sheets contain four additional occultations of Regulus.

Apart from its Aldebaran occultations, sheet 6 also counts 29 lunar eclipses. The mean of the absolute initial table errors of these lunar eclipses

---

[35] Cf. the paucity of observations in Cod. μ_6 made before the installation of the quadrant.

[36] There are only two exceptions: one is a *lapsus calami* of Oct. 5 for Oct. 25; the other is a date (1733 May 13) which occurs in the Cod. $\mu_{41}^{\sharp}$ page without being used, and which is omitted from the Cod. $\mu_{33}^{\sharp}$ sheet completely.

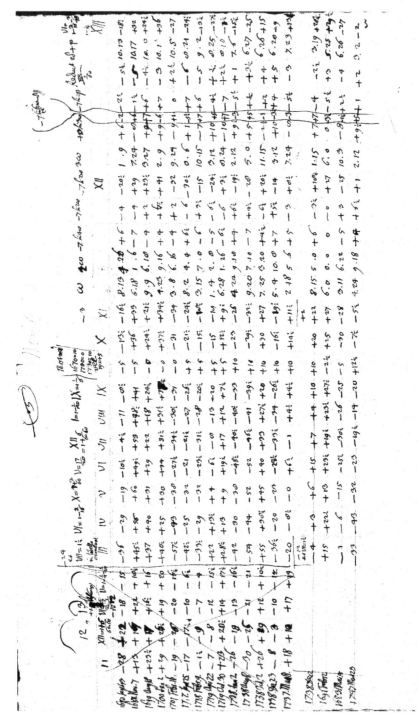

Fig. 8.3 Cod. $\mu_{33}^{\sharp}$, spreadsheet 7

is about $33''$, whereas the mean of the absolute errors of the solar eclipses in other sheets is $23''$. The poor result in case of the lunar eclipses may explain why they played only a minor role in the improvement of the tables. The disappearance of lunar features into the shadow of the earth during lunar eclipses is hard to time, because the gradual transition between umbra and penumbra, augmented by scattering in the terrestrial atmosphere, causes the earth to cast a fuzzy shadow on the moon. For the same reason, longitude determination from lunar eclipses tends to yield rather poor results.

Sheet 1. On sheet 1 we find first the same 14 solar eclipses as on sheet 8, followed by 21 occultations of Aldebaran including those already mentioned, then the four Regulus occultations, and finally 15 other occultations and conjunctions with mainly Alcyone, Antares, and Spica. Lunar eclipses are significantly absent from this sheet. As far as the data on this sheet occur on other spreadsheets, they have identical initial errors or at most a few seconds difference.

Tab.I/II. Two of the spreadsheets of Cod. $\mu_{28}^{\flat}$, marked 'Tab.I' and 'Tab.II', contain almost the same data set as on sheet 1, with identical initial table errors. Additionally they contain a selection of the lunar eclipses of sheet 6, also with identical initial errors. Sheet 1 was made difficult to read because Mayer had wedged in many data between already filled lines, but on Tab.I and Tab.II the inserted data had found their deserved slot.

Sheets 2–5. The remaining spreadsheets of interest are those numbered 2–5 of Cod. $\mu_{33}^{\sharp}$. These sheets all contain the same dates in largely the same ordering and with identical initial errors, save very few differences here and there. The dates on these sheets are mostly those on Tab.I/II augmented by about 30 occultations including a dozen of Regulus. The position calculations of these Regulus occultations are conspicuously absent in Cod. $\mu_{41}^{\sharp}$, in contrast with almost all other data. The initial tabular errors on these four spreadsheets are, as far as the dates allow comparison, considerably better than on Tab.I/II.

Sheet 3 has the peculiarity of an empty space where sheets 2, 4, and 5 list the 14 lunar eclipses. A mistake on sheet 5 that has been corrected on sheet 2 suggests that sheet 5 was made before sheet 2. In accord with this, sheets 2 and 4 have two dates in 1754 which are more recent than any of the dates on the other sheets. So sheets 3 and 5 must antedate sheets 2 and 4.

8.5.3 Analysis

Now that we have an idea of the dates and initial errors that went into the sheets, let us turn our attention to the content of the remaining columns. In order better to understand the relations between the spreadsheets and related manuscripts, and to test our interpretation, I decided to make an electronic

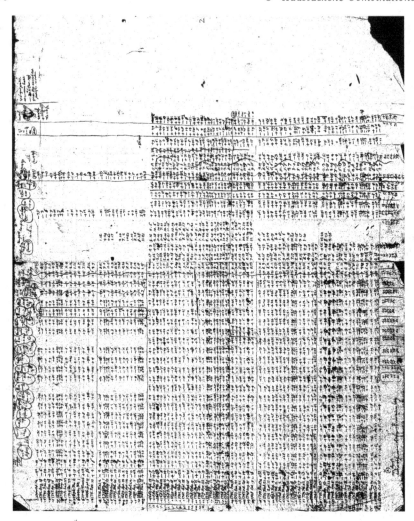

Fig. 8.4 Cod. $\mu_{33}^{\sharp}$, spreadsheet 2

version of them: indeed, using a spreadsheet program on a computer. First, we discuss some of the details of constructing the electronic mimics, and then we take a look at what conclusions they corroborate.

Ideally, the electronic spreadsheets fulfil a number of prerequisites. They assist in the interpretation of their original paper counterparts. Of course the lay-out of the paper and electronic incarnations must be the same. As a further requirement, it must be easy to feed the electronic spreadsheet with a different interpretation of a column in the paper spreadsheet, thus testing different possibilities in case of doubt. As a bonus we expect to spot some of

Mayer's calculation mistakes. Finally, the electronic version easily enables us to do some statistical analysis of the data, which then provides us with an idea of the effectiveness of Mayer's work.

An electronic spreadsheet is a matrix, consisting of rows and columns, whose elements can hold various kinds of data, including text, numbers, and functions. Looking at a spreadsheet on-screen, we see the text, the numbers, but usually not the functions: instead, we see the values that the functions produce, given the input which usually consists of the values in specific other locations of the spreadsheet.

Looking at Mayer's paper spreadsheets, we see rows and columns filled with numbers. The rows commence with a date, such as (taking the first data row in Fig. 8.3 as an example again) 1680 Sep. 13. The columns are headed by a brief description of their meaning: for example, '$-24''$ et VII $= 1 + \frac{1}{2}$' indicating a constant shift of $-24''$ and an increase of the seventh equation by half its value. The number below this column heading and in the row starting with the date 1680 Sep. 13 is the resulting error between predicted and observed lunar position, for that date, if the tables are modified as stated together with all the amendments specified in previous columns. Or, at least, in most cases it is; sometimes Mayer rejected some amendments and returned to a state in a former column.

We can safely assume that Mayer operated in the following way, exemplified for the example date and example amendment. He looked up the relevant position calculation for the event of this date (1680 Sep. 13), and inspected that calculation to take out the value of the VIIth equation, which is $-54''$. Half of this, or $-27''$, he then added to the error value in the preceding column of his spreadsheet, so he obtained $+15'' - 24'' - 27'' = -36''$ as the new cumulative error value.[37]

We follow the same strategy in the electronic spreadsheet, but we replace the looking up of the original equation value by a direct computation, as follows. We know not only the listed date of the event, but (through inspection of Mayer's position calculation) also the approximate time of day. For that instance in time we can compute the mean arguments of lunar longitude, apogee, and node using Mayer's mean motion coefficients (e.g., those of kil), and also the solar longitude and apogee. These five values can be computed once and for all for the complete collection of dates and times that appear in the spreadsheets. Thereafter we have them ready for the computation of any equation needed, since their arguments are just linear combinations of them. In fact, Display 8.1 contains a subset of these givens.

---

[37] This particular example is taken from the position calculation on sheet 7, Cod. $\mu_{33}^{\sharp}$, where, as we have seen, column III applies the stated amendment of the VIIth equation together with a shift of $-24''$. It turns out that Mayer treated all equations as independent. In reality they are not completely independent because the equations were applied in steps, and each step depended on the results of the previous step. Mayer justly disregarded the very small secondary effects of the amendments as they ripple through the steps.

Now, continuing our example, we are faced with the task of applying 'VII $= 1 + \frac{1}{2}$' for 1680 Sep. 13. Assuming that Mayer dealt with his kil tables, we guess that the VIIth equAtion is $+58'' \sin(2\delta - p)$; we may have to make a different guess in case the numbers disprove our assumption. We compute the argument $2\delta - p$ using the lunar mean arguments stored for this event, and find 293°. We substitute this in the supposed formula of the VIIth equAtion, and obtain $\frac{1}{2} \cdot 58'' \sin(293°) = -27''$. We can discern that our result agrees with his result, because it allows us to compute exactly the same value for the new column as we find in the manuscript. So perhaps our guess was right. Hoping for more successes, we repeat the same operation for the other dates down the same column, and if our results are generally in agreement with Mayer's, then we conclude that we interpreted 'VII $= 1 + \frac{1}{2}$' correctly. We should not be disturbed by finding small differences of $1''$ or $2''$, which may be easily caused by rounding errors. Larger errors do sometimes occur and if they are only few, they may be due to an error on the part of Mayer, but if there are many, then our interpretation needs revision. One relatively often occurring mistake of Mayer's is that he accidentally reversed the sign of the amendment (continuing the example, he would then add $27''$ instead of subtracting it).

In passing, we note that the prescription 'VII $= 1 + \frac{1}{2}$' is apt for Mayer's proposed method: it clearly expresses what to do with the looked-up value of the equAtion in the position calculation, without recourse to its argument. An alternative notation stating the number of seconds, such as VII $+28''$, would mean much harder work, but nevertheless there are some instances of it. Mayer also extended his notation to, e.g., the self-explanatory 'VII $= 1 + \frac{1}{2} - \frac{1}{20}$'.

It takes a while to implement all the relevant spreadsheets in electronic form, backing up to try a different interpretation when results do not match. The results are represented in Appendix B. For every sheet it shows a list of the successive amendments to the equAtions, and the effect of each amendment is shown in a box plot (as explained there). The appendix concludes with a display showing all amendments in all sheets.

Display 8.2, to which the following remarks apply, shows what proportion of the paper spreadsheets is confidently understood. Under 'columns' is a count of the columns in the sheets in which Mayer actually implemented an experimental amendment to an equAtion. Columns of convenience that just record an intermediate value of some kind are excluded from this count. The columns in the mimic with a high number of problematic results, indicating a lack of understanding on my part, are counted under the next heading. Under 'total' is the total count of all Mayer's coefficient amendments for all dates: the example above for argument VII on 1680 Sep. 13 counts as 1; since not all columns are completely filled in for all the dates on the sheets, the count is less than the product of the number of dates and the number of columns. Finally, under 'diffs' is a count of the cells where the reconstruction and the original differed by more than $5''$, originating either from built up rounding differences

Display 8.2 Statistics of the spreadsheet mimic (see the text for explanations of column headings)

Sheet	Columns	Problematic	Total	Diffs
Cod. $\mu_{41}^{\sharp}$ 44	10	0	130	1
47	16	0	288	8
342	16	0	214	2
Cod. $\mu_{28}^{\sharp}$ I	29	0	1,932	34
Cod. $\mu_{33}^{\sharp}$ 1	24	2	1,253	13
2	30	8	2,448	42
3	20	0	984	6
4	7	1	631	5
5	27	0	1,780	26
6	7	0	189	0
7	16	1	272	7
8	21	0	294	5
All	223	12	10,415	149

or from an error by either Mayer or me. This count is limited to those in the well-understood amendment-implementing columns. Whenever I discovered such a difference I adopted the value in Mayer's spreadsheet also for the electronic spreadsheet, to keep original and mimic as similar as practicable.

The listed counts show that I was able to understand about 90–95% of the sheets, certainly enough to base significant conclusions on.

8.6 Results

The 13 spreadsheets analysed in Sect. 8.5 were all made between the spring of 1753 (the publication of the kil tables) and October 1754. Thus, Mayer computed those more than 10,000 spreadsheet entries in at most 18 months. On average, he filled more than 550 entries a month, and this must have taken a considerable share of his time. Indeed, Mayer wrote to Euler on 6 March 1754 that he had applied most of his spare time to the improvement of lunar theory.[38] The effort that Mayer put into the improvement of his lunar tables was enormous and I have not been able to find a similar effort of model fitting in or before the 1750s on an even slightly comparable scale. We will now draw some conclusions from the analysis of the sheets.

First of all, let us return to Mayer's own remarks on his procedure as he wrote to Euler on 6 March 1754, discussed above on p. 146. We may now recognize several aspects that he mentioned there, such as: the central role for occultations, especially those of Aldebaran; different adjustments for

---

[38] Forbes (1971a, p. 80). As a side note, we remark that the intensity of his labour may explain, to a certain extent, the scarcity of signposts with which Mayer kept track of his own work; clearly he was so absorbed in his task that he was not overly anxious to get lost.

the epochs of differing bodies, reflecting catalogue errors of the stars; the presence of many 'although almost inappreciably small' corrections; and the total absence of lunar meridian transit observations. All these aspects are confirmed in the spreadsheet analysis.

The dates that Mayer employed, combined with the initial table errors, suggest a certain grouping among the sheets. Fol. 342 (on solar eclipses) is, most likely, the oldest, because it seems as if Mayer's technique is in a more rudimentary state compared to the other sheets. Sheet 8 is clearly connected with it, being the only other one that deals exclusively with solar eclipses.

The Aldebaran series of sheets consists of fol. 44 and 47, sheet 7. Perhaps sheet 6 (lunar eclipses followed by Aldebaran occultations) antedates fol. 44. With sheet 1 begins the phase where solar eclipses and star occultations are studied simultaneously. Lunar eclipses are also blended in on later sheets.

The next trials took place on sheets Tab.I/II and Mayer apparently reached a conclusion, because they resulted in a new set of table coefficients. Consequently, the kil tables must have been superseded by new ones, of the vlije, vlakte, and mui variety. This is evident when the changes made on Tab.I/II are compared to the differences of coefficients between kil and the latter group of tables.

The remaining sheets 2–5 start with a substantially better initial error column that was calculated from this fresh version of lunar tables. These sheets contain no amendments larger than $16''$ (except in the epoch of lunar apogee), contrary to all the sheets discussed above. These sheets try to improve the newer variety of the vlije, vlakte, mui, and similar tables.[39] Previously, we have seen that sheets 3 and 5 antedate sheets 2 and 4.

---

[39] The motivation for this goes as follows. First, when Mayer constructed sheets 3 and 5 he had decided to drop the VIIth equation (argument $2\delta - p$) and renumber the ones following it. This is consistent only with the val, vlije, and vlakte tables found in Cod. $\mu_{49}^{\sharp}$ and with mui. Vlakte is also consistent with an explicit reference to the main evection coefficient written on sheet 3. The equation with argument $2\delta - p$ reappears as the *last* equation before the reduction to the ecliptic.

Second, when an amendment is stated as a fractional change to the original equation, then its effect in arc-seconds puts limits on the value of the coefficient of that original equation (this is comparable to the extraction of coefficients from position calculations explained above). Sometimes these limits are narrow enough to restrict the table versions that can interact with the spreadsheet. With this principle it can be shown that only Cod. $\mu_{49}^{\sharp}$-versions and mui (which is very similar) match sheet 2.

Third, when we modify equations I to IX of kil using the amendments of sheet 1, the result matches the vlije version pretty well, except for the VIIIth: Kil's coefficient $+40''$ reduced by $\frac{1}{8}$ gives $+35''$, whereas vlije has $+30''$. None of the other table versions matches nearly as well. We have no indication of the major vlije coefficients for equations X to XIII. The amendments for those equations in sheet 1 (or on any of the spreadsheets that I presume precede it) are partially matched by the vlakte coefficients. The amendments in the remaining sheets 2–5 provide a weak link between the vlije-vlakte coefficients and val.

Finally, fol. 20v of Cod. $\mu_{49}^{\sharp}$ lists conclusions that are based on sheet 2. I could not determine if indeed Mayer prepared tables according to the amendments that he summarized there. The amendments are mostly, but not completely, consistent with a possible transition from vlije and val (the latter filling the gaps that the former leaves open) to

These points together demonstrate that Mayer used the early spreadsheets up to sheet 1 to improve the coefficients of the printed kil tables, probably resulting in the vlije tables of Cod. $\mu_{49}^{\sharp}$. From there, using updated initial error values, he started a new round of improvements on sheets 3, 5 and 2, 4 which also precipitated in the Cod. $\mu_{49}^{\sharp}$ tables, although it is less clear how. These results justify the interpretation of Mayer's process as depicted in Fig. 8.2.

8.6.1 Reflection

In the final diagram in Appendix B it can be seen that Mayer tried to fit almost every equation in his tables. The reduction (the XIVth) is the only exception.[40] He also made amendments to longitude (partly to correct errors in the catalogued positions of the occulted stars), mean motion, and apogee, and he tried several 'new' equations. No amendments were made to equations of latitude and parallax of the moon, however.

Working with the sheets, Mayer had to make decisions continuously. Time and again, he had to decide which equation to amend next and by how much. He needed to select successful amendments and to drop unsuccessful ones. And he had to decide when to make new tables based on the revised coefficients, going full circle in our conceived diagram. I have no indications that he made those decisions methodically. The spreadsheets formed a tool, and an appealing tool at that, but they do not embody a method. They would be part of a method when the whole process was formalized, including a mechanism repeatedly to select the next modification of yet another equation, and including a criterion to decide when to make new tables. Without these conditions, application of the process was confined to the individual of Tobias Mayer performing the specific task of improving his lunar tables. To speak of a method would require a generalization breaking at least one of these restrictions. In that respect, Gauss's verdict of *hausbackene Combinationen* was correct.

---

vlakte. The situation is rather complex and it seems that Mayer replaced some tables in some versions by new ones.

[40] As has been mentioned earlier, the reduction is a pseudoequation, it reduces the computed longitude of the moon in its orbit to the ecliptic. As such it is not a perturbation of the lunar motion, but it reflects a change of coordinate system. It depends on the inclination of the lunar orbit with respect to the ecliptic. Interestingly, Mayer remarked (Mayer 1753b, p. 387) that the inclination is variable due to solar attraction, but he took its mean value for the reduction, and worked the variable part into the variation. Although there are no adjustments to the coefficient of this equation in the spreadsheets, Mayer *did* revise its value from $-6'57''$ in the printed kil tables to $-6'51''$ in vlakte, vlije, and *Theoria Lunae*, and to $-6'44''$ in val and veen. These changes might reflect revisions of the mean inclination of the lunar orbit.

Although Mayer usually tried several different amendments to an equation, he was seldom in doubt regarding the *sign* of the amendment. That is, he had a pretty firm idea whether to increase or decrease any equation. I suggest that he was guided by visual inspection of equation values for the different dates in relation to the error values. To make successful decisions he would need a mind that is able to quickly assess large streams of numbers, which is perhaps a likely attribute for anyone successful in lunar theory. But although his proposed sign of an amendment was usually right, he sometimes computed the consequences of the amendment with an accidentally reversed sign.

The statistical analysis of the sheets, depicted in the appendix, show that Mayer was able to improve his tables significantly, to the extent that they could match existing observations. My analysis indicates that the standard deviations of the errors went down from $\sigma \approx 40''$ starting with the kil tables, to $\sigma \approx 15''$ on sheet 2. This would suggest that his declared goal to achieve tables accurate to $30''$, was reached in 95% of cases, at least when we are willing to assume a normal distribution of the errors. Lacking the modern concepts of standard deviation and normal distribution, Mayer supposedly meant to bring *all* errors down below $30''$, not just 95% of them or any other convenient percentage. Likewise, the requirements of the Longitude Act were stated in absolute, not statistical, terms. In connection with this, it is interesting to glance over the box plots in Appendix B and observe that Mayer was often more effective in reducing the outliers than the interquartile range (i.e., the length of the middle box) of his data set.[41] He may have reached further improvements in the other spreadsheets that I did not analyse. In order to appreciate the result of Mayer's work, I produced a least-squares fit of the table coefficients to the same data set as Mayer had employed in his spreadsheets. The modern fit accomplishes a standard deviation of $14.5''$, showing that Mayer's fit was remarkably successful: his 'homely combinations' nearly produced the best possible tables.

---

[41] The standard deviation for kil just mentioned is roughly in agreement with the one found in Display 6.3, but the final standard deviation of $\sigma = 15''$ on sheet 2 seems too optimistic when compared to $\sigma = 30''$ for the final rede tables. Presumably this is a result of either the totally different reference data sets (real observations in Mayer's case, against virtually error-free positions computed from modern ephemerides in Display 6.3), or else of the circumstance that Mayer's improvements in mean motion parameters are not taken in in the latter display.

Chapter 9
Further Aspects of Model Fitting

The previous chapter concerned Mayer's work of model fitting, as we would call it nowadays. Mayer had to deal with errors in his model, in its coefficients, and also in the observations that he employed to fit it. In fact, Mayer's work abundantly shows his ambition to take control over errors, and to increase accuracy and precision: be it in cartography, the mapping of the moon's surface, or the lunar orbit, or the consistent description of colours. It is evident in his design of instruments, in the 'repeating principle' that he introduced in angle measurements, in the *Mappa Critica*; it is the unifying theme in his work, as Forbes has already pointed out. 'The science of practical errors is so far not yet sufficiently developed',[1] wrote Mayer when setting out on an investigation into the limits of the human visual acuity under various light conditions – incidentally, that subject had cought his attention because of its impact on the accuracy of angular measurements.

We continue now with some aspects of Mayer's work where he shows his commitment to getting the best possible result out of the available data. We will gain some insight into the statistical aspects of his work. Mayer developed ad-hoc procedures in each case; the statistical tools of today had not yet been developed.

First, we investigate the quality of the observations at Mayer's disposal, particularly those which he used in the spreadsheets. This leads us to the realization that the result of his fit is almost optimal, but that the precision to which he worked was not matched by the accuracy of the data. Next, we briefly discuss a memoir of Mayer's in which he proposes a mathematical model of world temperature distribution. I argue that this memoir bears testimony to Mayer's trust in his spreadsheet tool. Then we turn to older work of Mayer's. We investigate his use of the arithmetic mean and his awareness of the cancelling property of random errors. Section 9.4 details another of Mayer's attempts to fit a model to observations: here we discuss his fit of a

---

[1] *'Scientia errorum practicorum nondum satis hactenus exculta'* (Mayer 1755, p. 120).

S.A. Wepster, *Between Theory and Observations*, Sources and Studies in the History of Mathematics and Physical Sciences, DOI 10.1007/978-1-4419-1314-2_9, © Springer Science+Business Media, LLC 2010

model of lunar libration to data via a redundant system of linear equations. Section 9.5 is devoted to the role of redundant systems of linear equations in the shaping of his lunar tables.

9.1 The Quality of Observations

As we have seen (particularly in the previous chapter), Mayer went to great lengths in fitting the lunar tables to observations. An interesting question concerns the quality of the observational data on which the fit was based, more specifically: how much random error is present in the data, and to what extent was Mayer aware of that? To investigate these questions, I sampled the position calculations in Cod. $\mu_{41}^{\sharp}$ between fol. 31r and 72v, which are all related to occultations of Aldebaran. Mayer used almost all of these in his spreadsheets, and rejected only those that seemed to yield very unlikely results. These occultations allow us to obtain a unique impression of the quality of data at Mayer's disposal, as I will explain next. The result will enable us to see Mayer's fitted tables in a new quantitative perspective.

An occultation of a star by the moon provides two sharply defined observable phenomena: the disappearance of a star behind the disc of the moon (called its *immersion*), and its subsequent reappearance (or *emersion*). The observation of an occultation consists mainly of the recording of the (local) times of these phenomena. When due consideration is given to the lunar parallax and diameter, and to the location of the point of contact on the lunar disc relative to its direction of motion, the observations provide very accurate positions of the moon relative to the star. This property makes occultations particularly suitable for Mayer's purpose.

The temporal separation between the two phenomena can never be much longer than an hour, because in that time interval, the moon appears to move approximately its own diameter relative to the stars. Such a time-span is very short compared with the periods of the lunar inequalities. Therefore, the errors of the periodic equations in the lunar tables before and after an occultation are equal for all practical purposes. Mayer already had the mean motions approximately right, so these too will produce equal errors before and after the occultation. Thus, the predicted lunar position will differ from its true position just as much at immersion as at emersion. This property makes *pairs* of occultation observations particularly suitable (apart from their usefulness to Mayer) to gain insight in the observational errors.

Mayer had a collection of occultation observations from his century and the previous one, made by several observers, and from various locations. These were the most important data that he applied in the spreadsheets to improve the coefficients of his tables. We will now use the immersion–emersion pairs of Mayer's position calculations to pursue our own aim, referring to the example calculation in Fig. 8.1. The example concerns an Aldebaran occultation on 2 January 1738 observed from Paris. It occupies two facing folios, the left side for the immersion, and the right side for the emersion.

In that example, Mayer computes the lunar position from the kil tables, following the general scheme set out in Sect. 4.1. This computation occupies approximately the top one third of both folios. Below that, Mayer reduces the observations, allowing for aberration and parallax (but not for nutation).[2] Apparently, he has two different observations of immersion and also two of emersion.[3] Consequently, he derives four errors, i.e., differences between computed and observed positions of the moon: for immersion, they are $-33''$ and $-21''$, and for emersion, he finds $-20''$ and $-28''$. He concludes that the average error is $-25''$.[4] This averaged result is a measure of agreement between observations and theory for the specified instance, which of course is precisely why Mayer had made these calculations.

But to investigate the data quality, our attention is now drawn to the dispersion of the errors around the average: it shows us the relative quality of two different observers, each observing one immersion and one emersion. Observer number one obtained a spread of $|33'' - 20''| = 13''$, but number two had a sharper result of $|21'' - 28''| = 7''$. These spreads are independent of the quality of the lunar tables, because the error of the tables is the same at the beginning and end of the occultations. But they do reflect errors in the observations and their reduction.

By taking notice of Mayer's derived errors at 34 pairs of im- and emersion, spanning the period from 1680 to 1750, I found a spread of $10''$ or less in 20 cases from $10''$ to $20''$ in 5 cases, and the remaining 9 cases showed a spread between $30''$ and $69''$.[5] It is illustrative to compare these numbers

---

[2] The fast change of parallax in right ascension provides the reason why it is not possible to average the times of immersion and emersion and make a position calculation for the middle of the occultation only, thus saving half the work. The effects of aberration and nutation can add about $20''$ and $17''$, respectively, to lunar longitude; however, since they are slow-changing, they have no impact on the sequel of this paragraph. (Aberration is an apparent deflection of light rays, resulting from the finite speeds of light and of the earth.).

[3] In Cod. μ_{12}, where Mayer collected his data, there are two references to this occultation: on fol. 63v, copied from the Mémoires for 1739, and on fol. 80v referring to correspondence between Lowitz and P.C. Maire. He attributed the second observation to Lemonnier.

[4] The arithmetic mean of all four results is $-25\frac{1}{2}''$.

[5] Taking half of the stated difference between immersion and emersion as the observation error (which is admittedly a best-case scenario), we see that Mayer's own estimate of $5''$ to $10''$ for these observations is slightly optimistic ('We have a considerable number of observed occultations of Aldebaran; out of these I have calculated the positions of the Moon with the help of the parallax, so that there can be no error of $5''$ or $10''$ in them' (Forbes 1971a, p. 81)).

The quality of the observations, which I have just related to the difference of errors before and after the occultation, depends mostly on the ability of the observer to trap the exact instance of the star's disappearance and reappearance at the limb of the moon: a longitude error of $10''$ is equivalent to a clock error of about 20 s. Before immersion, the observer can see the star approaching the limb, whereas the emergence occurs without warning; this makes the immersion (in principle) easier to observe. Other factors that play

to the standard deviations that I computed for Mayer's spreadsheet results as summarized in Appendix B. Mayer accomplished to bring the standard deviation down to 15″, very close to the least-squares fit of 14.5″. Our current investigation shows that the quality of his fitted tables matches the quality of his data. This means that Mayer could not have achieved a significantly better result than he did.

Although the end result that Mayer achieved is optimal, he exerted himself excessively to reach it. Page after page, occultation after occultation, he must have noted the dispersion in pairs of observations. Seeing such a dispersion, usually in the order of 10″, did not restrain him from filling in his spreadsheets to a precision of half a second. For us, this is hard to understand. 'Unwarranted number of significant digits', we tend to remark. We encountered another example of this when we studied the relation between the single-step and multistep equations in *Theoria Lunae*, on p. 139. Donald Sadler observed the same phenomenon in the early Nautical Almanacs and in Nevil Maskelyne's procedures for the computation of (geographical) longitude by lunar distances and the accompanying tables. With one significant digit less (i.e., with minutes and seconds replaced by minutes and tenths of minutes), the resulting longitude would have suffered inappreciably, but the burden on the human computer would have decreased considerably, mainly because a number of corrections could then be skipped on account of their small size.[6] Certainly numerous other examples exist of this phenomenon of over-precision in the eighteenth century. It seems to me that an attitude towards data prevailed in which a *lack of accuracy* (such as the dispersion in the input data at Mayer's disposal) was accepted, by many even recognized as partly inevitable; the self-cancelling property of random errors was recognized by many. But at the same time, we see both Maskelyne and Mayer on their guard against any *loss of precision* without realizing the futile part of their effort in view of the limited accuracy of the data.

---

a role include the observer's experience, the meteorological circumstances, and whether the moon is waxing or waning (i.e., which of the moon's limbs is illuminated).

Incidentally, the quality of the data at Mayer's disposal (as here investigated from pairs of observations) matches the quality that Tycho Brahe achieved in his determination of the longitude of α Arietis, after he had combined his data in pairs to eliminate certain systematic errors. This conclusion shows clearly the unprecedented accuracy attained by Tycho, a conclusion that is only slightly moderated when it is taken into account that Tycho built his data set personally and for the specific purpose of fixing the longitude of α Arietis as a reference star. For details on Tycho's determination, see Hald (1990, pp. 145–146) and Plackett (1958).

[6] Sadler (1977, pp. 115–119).

9.2 World Temperature Distribution

The careful treatment of astronomical observations roused Mayer's interest in the refraction of light rays through the earth's atmosphere. The amount of refraction depends on the density of the air and hence on the temperature. It is for this reason that he came to consider, in the memoir *A More Accurate Definition of the Variations of a Thermometer*,[7] a mathematical model of the temperature of places on earth, dependent on various geographical data. Mayer specifies a formula for the average annual temperature of a place as a function of the latitude. Then he provides several refinements in the form of: a correction term for the altitude, a term for annual (seasonal) variation, and a term for diurnal variation. Mayer asserts that particular attention must be given to the period, phase, and amplitude of the last two variations. These are governed by certain coefficients in the formulae, whose values are quite easily obtained with the help of observations. Mayer gives only a few simplified examples of the determination of those coefficients, but no thorough treatment of the problem. He does not claim any predictive power for his model. Rather, his main goal is to propose a method to investigate the actual temperature data: 'I think that it is impossible to define the causes and quantities of the remaining, more involved, variations, unless the effects of the different causes are analysed in the way which I have roughly sketched out here.'[8]

Forbes and Delambre have commented that the interest of this memoir by Mayer lies more in the method than in the results.[9] Mayer himself had put forward that astronomers have good methods to investigate the movements of the luminaries, and that meteorologists have something to learn from astronomers:

> Therefore, at this point, we may transfer the example of that astronomical method to variations in the atmosphere, applying it in particular to the ratio of heat and cold: it will thereby be possible to show how I deem that meteorological observations ought to be treated, so that richer fruits can be expected from them.[10]

The relation between Mayer's memoir and his interest in correcting raw astronomical observations for atmospheric refraction has already been noted by Forbes.[11] Forbes summarizes Mayer's proposed method (in the *Thermometer* treatise) as 'isolating a major periodicity, examining the residuals for a second-order periodicity, etc., until the observations had been analysed into a series of independent periodic functions each characterized by a mean

---

[7] Original Latin text posthumously published in Mayer (1775); translated and republished in Forbes (1971c).

[8] Forbes (1971c, p. 61).

[9] Forbes (1971c, p. 21), Delambre (1827, p. 447).

[10] Forbes (1971c, p. 54).

[11] Forbes (1980, pp. 178–181).

value and its variation about the mean'[12] and concludes that the relationship to Mayer's lunar investigations is obvious, aiming at the analogy of periodic fluctuations, and the customary way among astronomers to investigate orbital motions. However, a major difference between the models, not pointed out by Forbes, is that the lunar theory was already able to supply a priori approximate values for the amplitudes of the anticipated periodicities, a feat which was (in Mayer's era certainly) unmatched by the theory of heat.

We note that Mayer had such a strong confidence in his astronomical perturbation analysis techniques, that he did not hesitate to export them to other disciplines. Significantly, he read the *Temperature* memoir to the Göttingen Scientific Society on 13 September 1755, when he had already developed the spreadsheet tool and when his *Theoria Lunae* was just about finished. Certainly, Mayer's approach bears witness to both his interest and his confidence in fitting models with many periodic terms. With some imagination, we can picture Mayer studying spreadsheets of temperature data.

9.3 On Averaging and Cancelling

The rest of this chapter is concerned with aspects of data use and model fitting present in Mayer's work prior to 1751, i.e., when he was still working with the Homann heirs in Nuremberg. Currently, we will investigate two related aspects of working with data: the fact that averaging over several data usually leads to a better result, and the property of random errors to cancel each other. A probabilistic proof that the arithmetic mean of a series of observations is more reliable than an individual observation was first given in 1755, by Thomas Simpson.[13]

In a memoir *Untersuchungen über die geographische Länge und Breite der Stadt Nürnberg*, Mayer evaluated older observations made by Wurzelbau in Nuremberg to determine the latitude of his habitat.[14] As Forbes mentioned,[15] Mayer's intention was to warn against the use of results (in this case, the latitude of Nuremberg) without a proper investigation into their origin (here, Wurzelbau's observations). I regard Mayer's memoir as illustrative of his use of data from other astronomers; he may even have started it to show his skill.

First, Mayer gave four series of Wurzelbau's observations: two series of superior culminations of Polaris, and two series of inferior culminations. He rejected the inferior culminations because the individual observations in

---

[12] Forbes (1971a, pp. 15–16).

[13] Plackett (1958, p. 124).

[14] It was published posthumously in Forbes (1972, I, pp. 33–44). The memoir deals only with the latitude of Nuremberg, and the title concedes that the memoir remained unfinished. An article in the Göttingen *Commentarii* Mayer (1752), also on the latitude of Nuremberg, is mostly concerned with Mayer's own observations.

[15] Forbes (1972, I, p. 9).

them deviated considerably more from the mean than those in the superior culminations. This shows that he considered the reliability of his data, and that he was prepared to select only the most reliable ones.

Next, Mayer averaged the remaining series, as was not uncommon among astronomers, and then computed Wurzelbau's latitude from the mean altitude of Polaris at upper culmination (he also used an independent determination of the polar distance of that star, which he could otherwise have derived from the combination of the superior and inferior conjunctions).

Mayer also explored two other series of observations that Wurzelbau made with different instruments, of culminations and solstices, but these comprised much shorter data series. Mayer used each of those other data series for a separate computation of the latitude of Nuremberg. These several different latitude determinations were apparently in reasonable agreement, at least after Mayer had corrected the results for errors in Wurzelbau's instruments, which he had detected through his careful analysis of the observations. Two of the three results differed by less than $2''$; the third differed from both by almost $20''$. Mayer felt justified to average over all three results, and then to round off towards the 'outlier', thereby implicitly giving it a slightly larger weight. We see that he was careful to cross-check results before he relied on them.

Yet there are certain differences between his procedure and modern ways of data handling. Mayer averages his computed latitudes without weighing them, although the data series from which the results were obtained are of unequal length and of unequal quality. So Mayer's end result is an average of several averages of unequal reliability. Instead, a modern statistician would prefer to compute latitudes from Wurzelbau's observations individually, and average only one time over the complete corpus, perhaps with weights assigned if the latitudes are of unequal reliability. Stigler points out that until the second half of the eighteenth century, astronomers were willing to average only among observations that were taken under comparable circumstances (same observer, same instrument, same object, etc.).[16] A procedure of taking weighted averages had already been outlined by Roger Cotes.[17] Mayer was evidently willing to average results obtained under unequal circumstances. But he did not give these results unequal weights.

We now turn to the cancelling property of random errors. While in Nuremberg, Mayer had started to write a treatise on map making, *Von der Construction der Land-Karten*, which was to remain unfinished. One of its topics was a discussion of the value of Roman itineraria for map making. Mayer remarked that those works usually specified distances in rounded Roman miles of 1,000 feet, without a fractional part, so the distances were not exact. But he added that sometimes these distances would be too large, and

---

[16] Stigler (1986, p. 30).

[17] See Gowing (1983, pp. 107–109).

sometimes too small, and he recognized that generally the rounding would have no appreciable effect on the end result. Thus, Mayer showed that he understood a basic property of random errors.[18]

The property that random errors tend to cancel each other is also applied in Mayer's design of two angle measuring instruments, the repeating circle and a modified recipiangle. Both these instruments are designed to accumulate a series of angle measurements for the observer. The repeating circle, for instance, allows the observer to measure an arc not only once but several times in succession, without intermediate readings of the instrument. The individual measurements are automatically summed on the circular scale of the instrument, and the observer has only to read the accumulated sum. After division by the number of observations, he obtains the average of his individual arc measurements. In this way not only random observer errors (such as faulty alignment of the telescopes and reading error) are reduced, but also the fixed and therefore systematic errors in scale division and collimation error. However, the observer can no longer recognize and discard outliers.[19]

At first sight, it might seem strange that the errors due to scale division are reduced in this way. Buchwald points out that Mayer (and Borda, who developed the instrument further) had in mind to reduce what we now call the systematic errors, in their instruments, not the random errors caused by the observer. Buchwald goes on to show that the instrument accomplishes just the opposite: that it fails in particular to average out the errors in scale division.[20] However, a circular scale of a given radius offers a fixed amount of space for a fixed number of divisions, so that the average division error must be zero. If, in the process of engraving, there is a systematic error which makes the divisions slightly too large on any part of the scale, then the engraver has to make up for the accumulated errors in another part. A bad workshop might produce a scale where almost every division is only slightly too large, compensated by a few that are way too small, but a careful craftsman would endeavour to avoid such errors. Therefore, the error distribution of the scale divisions may be assumed more or less symmetrical and independent identically distributed. The observer takes a series of n consecutive measurements and reads the scale only before and after the complete series. Let δ be the error due to the scale divisions and note that, on a good instrument, δ is independent of n. Then the averaged result has an error of δ/n. In contrast, n individual observations taken with an octant or a sextant involve n readings each on the same part of the scale with the same systematic errors. Thus, repetition of the observation on an instrument with a circular scale indeed reduces the scale division error.

---

[18] Mayer's text has the full title *Von der Construction der Land-Karten mit dem Exempel einer Karte von Ober-Teutschland erkläret* and has been republished in Forbes (1972, I, p. 49).

[19] The recipiangle is an other angle measuring instrument, used in surveying, which will not be further discussed here. See Forbes (1980, pp. 153–154, 158–164).

[20] Buchwald (2006, p. 568). The current discussion has benefited considerably from an interesting exchange of ideas with him.

Mayer advocated that his repeating procedure

[...] is of such a nature that it [...] relieves the smallness of the instrument, and sets the errors right that arise from the uncertainty of the divisions. For I suppose that it is hardly possible to distinguish 2 or 3 minutes due to the slipperiness of the divisions (although single minutes could perhaps be distinguished with the help of a micrometer) and I have shown that by taking the observations in sixfold with the special rule, this same error can be depressed to 30''. This manner differs altogether from that which mathematicians apply in other cases, since they are accustomed to take the arithmetic mean between many observations of angles, but not of a continuous series [of observations] captured on an instrument; for in that ordinary manner, the error is only diminished somewhat, while in this other [manner] it is deminished necessarily.[21]

Earlier in his description he had suggested that the rim of the repeating circle be graded in parts of 15'; an observer would then indeed be able to estimate his reading to a precision of 2' or 3'. Mayer makes no clear distinction between (random) reading error and (systematic) division error, and perhaps such a distinction is immaterial here. The point is that no inaccurate readings from an inaccurate scale are necessary while collecting data.[22]

9.4 Libration: A Case of Model Fitting

Any investigation of Mayer's work and model fitting has to take into account Mayer's lucidly written tract on the rotation of the moon *Abhandlung über die Umwalzung des Monds*.[23] Although Mayer himself was perhaps more interested in the direct results that he had obtained there, his work attracted the attention of many for his original approach to the much more general topic of the combination of observations, which in this particular case resulted in an overdetermined system of equations for three parameters.

This work of Mayer's, which we will now discuss, was reviewed by Lalande[24] and generalized into a method by Laplace; the latter's generalization became widely known as 'Mayer's method' and as such it was used during the first half of the nineteenth century, among others by Laplace's

---

[21] '[...] ea ejus natura est, ut parvitati instrumenti [...] subveniat, corrigatque errores ex divisionum incertitudine oriundos. Suppono autem duo vel tria minuta propter hanc lubricitatem divisionum distingui vix posse, etsi forte singula micrometri ope liceret, et docui, quod sextuplicando peculiari ratione observationes idem error ad 30'' deprimatur. Differt autem omnino hic modus ab eo, quem alias adhibent mathematici, cum medium arithmeticum inter plures angulorum observationes, sed non continua serie in instrumentum exceptas, solent accipere; hoc enim vulgari modo error probabiliter tantum, isto vero necessario diminuitur' (Mayer 1770, p. 37).

[22] Mayer explicitly tried to control random operator error also in his design of another instrument, a new astrolabe for land surveying; see Forbes (1971b, p. 114).

[23] Mayer (1750a).

[24] Lalande (1764, 1st ed., pp. 1234–1243).

assistant Delambre.[25] It is quite likely that Mayer obtained his ideas from
a memoir by Euler.[26] Later researchers have assigned at least two other
names to it: *Method of Equations of Condition* and *Method of Averages*.[27]
The first is rather nondescript, and that is why I prefer the second name –
it catches the essence of the procedure, even though it is factually wrong
because it is not a method and Mayer did not take averages. Mayer used
neither of these names; indeed we may well ask whether he himself regarded
it as a method or rather as an ad-hoc procedure. He seems to have made
little or no use of it in other places, and it is less general than is perhaps
apparent at first sight. Although I prefer not to regard it as a method, I will
adhere to the conventional naming of method of averages.

After an introduction to the topic of Mayer's tract, and a technical
exposition of the geometry of the problem he endeavoured to resolve, we discuss his method of averages in Sect. 9.4.2. Quite naturally, the question arises
whether Mayer used this same method to adjust the coefficients of the tables
of lunar motion.[28] Above, I showed that he had a different procedure for that
purpose around 1752 and later; however, the two are not mutually exclusive.
To find a more definitive answer, I have looked specifically for places where
Mayer uses similar systems of equations. Some of these I discuss in Sect. 9.5.

The goal of Mayer's research on lunar libration was to improve the mapping of the visible lunar surface. This was important partly because of its
relation to the determination of geographical longitudes via the timing of
lunar eclipses, and partly because Mayer had set himself and the Homann
cartographic office annex Cosmographical Society where he was then working,
the prestigious task of producing accurate lunar globes.[29] Although Hevelius
and Riccioli had mapped the visible part of the surface of the moon, and
Jean-Dominique Cassini (I) had initiated an investigation of the libration
(discovered by Galilei and explained below), Mayer complained that the current state of selenography was rather poor: there was no consensus in the
nomenclature of lunar features, and Cassini's research was not up to the
attainable standards of accuracy.

First now, let us get involved with the problem at hand. It is well known
that the moon always turns the same side of its surface towards the earth.
Upon closer inspection this turns out to be only approximately true: for several reasons the moon is subject to a slight apparent wiggling called libration.

---

[25] Stigler (1986, p. 31).

[26] Lalande (1764, 1st ed., p. 1241); the Euler memoir is Euler (1749a). Also see the final
section of this chapter.

[27] The first of these names is used for instance by Forbes and Wilson (1995, p. 66), cited
below. In a different place in the same book, Schmeidler uses the latter name, reserving
the former for the equations that make up the overdetermined system (Schmeidler 1995,
p. 201).

[28] See the quote from Forbes and Wilson on page 149.

[29] See fn. 4 on p. 29.

The reasons for this wiggling are as follows. First, due to the diurnal motion of the terrestrial observer, his aspect of the moon varies between moonrise and moonset. Second, the moon rotates (practically) uniformly around its axis while its velocity of revolution around the earth varies: consequently, a terrestrial observer sees sometimes a bit more of the leading half of the moon's surface, and sometimes a bit more of the trailing half. Third, the rotational axis of the moon is not perpendicular to its orbital plane. As a consequence, we look upon the moon's north pole for half a month, and on its south pole during the other half. The fourth effect is the slightly perturbed rotation of the moon due to its deviation from perfect sphericity. This effect is much smaller than the other librations.

An accurate mapping of the lunar surface can only be arrived at when the measured positions of its features are corrected for these librations. The first two librations depend mainly on the parallax, longitude, and latitude of the moon, which are more or less observable. The third libration depends on the orientation of the lunar axis of rotation. It is this orientation, and Mayer's investigations of it, that will concern us here. The fourth libration is too small to be observable for Mayer: this so-called physical libration was first hypothesized by Newton, but only in the 1840s were sufficiently accurate measurements available to prove its real existence.[30]

9.4.1 Locating the Rotational Axis

In section 13 of his libration tract, Mayer sets out to deduce, from observations, the orientation of the lunar polar (or rotational) axis. First, he constructs an equation that models a relation between the orientation of the axis and certain observable quantities (see (9.2) below). Subsequently, he uses a surplus of data to derive the axis orientation as well as an error estimate. We start now with the derivation of the model equation; readers only interested in its application may skip to (9.2).

The geometry of the problem is illustrated in Fig. 9.1, which is taken from the original publication. The figure represents the moon with its centre in C. Let the moon rotate around an axis through CP, with P its north pole. Let the equator of the moon be the great circle $NLZnz$. Draw the great circle $YWNBXn$ parallel to the ecliptic; Mayer calls it the *lunar ecliptic*. Let point A be a pole of it.

---

[30] A non-spherical moon – or rather the hypothesis of it – had entered celestial mechanics in a different way, too. D'Alembert and Euler had tested, independently of each other, whether a non-spherical mass distribution of the moon could be held responsible for the missing part of the motion of the lunar apse line. They concluded that it could – but only if the moon were extremely dumb-bell shaped (Waff 1995, p. 40).

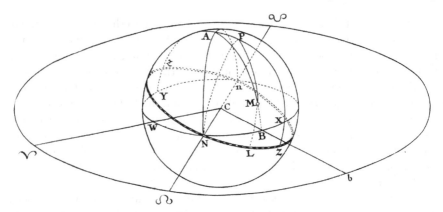

Fig. 9.1 Geometry of the lunar rotational axis problem

The rotational axis of the moon is not perpendicular to the ecliptic. Therefore, the planes of the lunar ecliptic and equator intersect in the line NCn. The points N and n on the lunar surface are called the *equinoctial points*. These points are not fixed relative to the surface of the moon; instead they both traverse the complete equator $NLZnz$ during one revolution of the moon about its axis.

The circle $\Upsilon \Omega b \mho$ is the lunar ecliptic projected onto the celestial sphere from the centre C. Mayer chooses, in accord with common practice, the direction of the vernal equinox $C\Upsilon$ as the zero point of ecliptic longitude. A little less common is his use of the symbol Ω, as I will now explain. Cassini had found in earlier research that the longitude of the equinoctial point N coincides with the longitude of the ascending node of the lunar orbit. Mayer comments that such is unlikely to be true all the time, because the motion of the orbital nodes is perturbed by the attraction of the sun while the axial rotation of the moon is uniform.[31] Still, Mayer chooses the symbol Ω (normally associated with the ascending node of the lunar orbit) to represent the point on the celestial sphere that corresponds with the equinoctial point N.

In the previous sections of his tract, Mayer had described how he had observed, between April 1748 and March 1749, 37 positions of Manilius, a distinct crater on the moon not far from the equator, represented by M in the figure. Of the 37 observations, Mayer had made a selection of the 27 most appropriate ones.[32] After extensive reduction, he deduced from these

---

[31] Except for the fourth form of libration mentioned in the introduction to this section. In a memoir of 1780, Lagrange concluded that the phenomenon discovered by Cassini, when taken in the mean sense, was a consequence of this physical libration (Wilson 1995c, p. 112).

[32] 'But to that end I have selected only those observations that are more correct than others and, because of the circumstance the moon was in at that time, more fit to the examination of the orientation of the lunar axis.' ("*Ich habe aber dazu nur diejenigen Beobachtungen ausgelesen, welche vor andern richtig und wegen der Umstände, in welchen sich der Mond*

27 observations equally many values for the following quantities: Manilius' *ecliptic* polar distance $h = AM$; Manilius' ecliptic longitude $g = \Upsilon b$; and the mean longitude of the orbital ascending node k, which is *approximately* the arc $\Upsilon \Omega$ in the figure (remember that Mayer's Ω here corresponds with the equinoctial point N). Actually Mayer took the value of k from existing lunar tables for the time of observation.[33]

Mayer's goal of finding the orientation of the lunar axis entails the determination of the angle $\alpha = AP$ between the ecliptic poles and rotational poles, and the *precise* longitude $\Upsilon \Omega$ of the equinoctial point N. He represents the latter angle by $k + \theta$, with k (already determined as explained above) the longitude of the ascending node, and θ (which is not in the figure) the unknown and constant small arc separating the ascending node from the equinoctial point. As we will see, he will also have to determine the latitude of Manilius $\beta = LM$. Mayer presents evidence (which I omit here) that the orientation of the lunar axis is fixed or, in the worst case, changing so slowly that it can be regarded as fixed over the time span of his observations.

To derive a relation between the known quantities g, h, and k, and the unknown parameters α, β, and θ, Mayer considers the spherical triangle APM. The cosine rule of spherical trigonometry, applied to this triangle, is

$$\cos PM = \cos AP \cos AM + \sin AP \sin AM \cos PAM.$$

Substituting $PM = 90° - \beta$, $\alpha = AP$, and $h = AM$, and noting that $\angle PAM = \angle NAP - \angle NAM = 90° - (g - (k + \theta))$, yields

$$\sin \beta = \cos \alpha \cos h + \sin \alpha \sin h \sin(g - k - \theta). \qquad (9.1)$$

Mayer says that, in principle, three observations suffice to solve for the three parameters α, β, θ, but that in practice this proves to be very difficult. Therefore, he approximates the last equation as follows.

In triangle APM, we have $MP - PA < AM < MP + PA$, or equivalently $90° - \beta - \alpha < h < 90° - \beta + \alpha$. By inspecting his data set, Mayer could deduce from the difference of the minimum and maximum values of h therein, that $\alpha \approx 1°40'$. With this preliminary value for α, he could estimate $\beta \approx 14°42'$ and $\theta \approx 10°$ or so. Therefore, he felt justified to take $\sin \alpha = \alpha$, $\cos \alpha = 1$, and $\sin \alpha \cos \theta = \alpha$. Expanding $\sin(g - k - \theta)$ and making approximations, the right-hand side of (9.1) yields $\cos h + \alpha \sin h(\sin(g - k) - \cos(g - k) \sin \theta)$. For the left-hand side, he introduced n satisfying $90° - \beta = h - n$, and

---

damals befunden hat, zur Untersuchung über die Lage der Mondaxe tauglicher sind, als die übrigen.") (Mayer 1750a, p. 122).

[33] These tables do not have to be of very high accuracy, since they here serve only to supply the *mean* longitude of the ascending node. The coefficients of its mean motion are computed easily and accurately from observations of eclipses one or more Saros periods apart. To find Manilius' longitude g, Mayer also needed the true ecliptic longitude of the moon, which he had to take from observations because the tables at his disposal were not yet accurate enough for that.

consequently $|n| < |\alpha|$; he then obtained $\sin \beta = \cos(h - n) \approx \cos h + n \sin h$. The two sides together give, after division by $\sin h \neq 0$,

$$n = \beta - (90° - h) = \alpha \sin(g - k) - \alpha \sin \theta \cos(g - k). \qquad (9.2)$$

This is a linear relation between the unknown parameters α, β, and $\sin \theta$, with numerical coefficients computed from the known angles g, h, and k. Values can still be expressed in degrees, the linearization of $\sin \alpha$ and $\sin n$ notwithstanding, because the linearization introduces the same factor $180/\pi$ on both sides of the equation. The neglected second-order terms are of the order of 10^{-4}. Modern least-squares solution of the system of equations (9.1) with Mayer's data differs from the solution of the linearized system (9.2) by only a few seconds in α and β, and about $4'$ in θ. This is too small to affect Mayer's conclusions presented below, because (as we will see) the estimated standard deviations of the three parameters are considerably larger.

9.4.2 The 'Method' of Averages

Now that we have this relation (9.2) between the observed and the unknown quantities, we will discuss Mayer's use of observations to find the unknowns. Mayer selected three observations with respectively a large, medium, and small value for h, to obtain three well separated equations in the unknowns α, β, θ:

$$\beta - 13° \, 5' = +0.9097\alpha - 0.4152\alpha \sin \theta,$$
$$\beta - 14°14' = +0.1302\alpha + 0.9915\alpha \sin \theta,$$
$$\beta - 15°56' = -0.8560\alpha + 0.5170\alpha \sin \theta.$$

Solving, he obtained $\alpha = 1°40'$, $\beta = 14°33'$, and $\theta = 3°36'$.

While in previous decades an astronomer would have been perfectly satisfied with this result and would have moved on to other business, Mayer remarked:

> But because errors are often to be supposed in the values of g and h that are deduced from observations, which [errors] are impossible to avoid, yet they have an influence on the values of α, β and θ: so also must we not completely trust the present determination, which is deduced from only three observations. One must only try three other observations to get convinced of this.[34]

---

[34] *"Weil aber in den aus den Beobachtungen geschlossenen Größen von g und h manchmal auch Fehler zu vermuthen sind, die sich unmöglich vermeiden lassen, gleichwol aber in die Größen von α, β und θ einen Einfluß haben können: So dörfen wir auch der gegenwärtigen Bestimmung, die nur aus dreyen Beobachtungen hergeleitet worden, nicht völlig trauen. Man darf nur eine Probe mit dreyen andern Beobachtungen anstellen, um hievon überzeugt zu werden"* (Mayer 1750a, p. 151).

To reduce the influence of the observational errors on the solution, Mayer's remedy is to take all his 27 observations simultaneously ('*zugleich*') into account, each yielding one of 27 near-linearized equations. To solve the resulting overdetermined system, he divided the equations in three classes of nine each. The first class held the nine equations with the largest positive values for $\sin(g - k)$, the second class those with the extreme negative values for $\sin(g - k)$, and the remaining equations (i.e., those with large values for $\cos(g - k)$) went into the third class. Then he summed the nine equations in each class to obtain again three equations in three unknowns, with deliberately large differences between the coefficients:

$$9\beta - 118° \ 8' = +8.4987\alpha - 0.7932\alpha \sin\theta,$$
$$9\beta - 140°17' = -6.1404\alpha + 1.7443\alpha \sin\theta,$$
$$9\beta - 127°32' = +2.7977\alpha + 7.9649\alpha \sin\theta.$$

from these, he obtained the solution $\alpha = 1°30'$, $\beta = 14°33'$, and $\theta = -3°45'$.

Here we have the key idea of the method of averages: divide the total corpus of equations in as many classes as there are unknowns, then sum the equations in each class, and solve the resulting system of equations. The solution is believed to be more accurate if more observations are employed, and if the division in classes is aimed at maximizing the differences between the coefficients of the final equations:

> [...] the advantage consists therein, that through the above division in three classes, the differences between the three sums become as large as possible. And the larger these differences are, the more correct are these unknown quantities α, β, θ to be found from them.[35]

In the case of this particular model, division in classes works surprisingly well to maximize the differences between the coefficients of the three summed equations. We will return to the reasons behind this later in the section. Mayer does not supply a reasoning why all classes should contain an equal number of equations.

Next, Mayer comes to a very interesting error estimation, based on the difference of the two solutions just quoted. Comparing the two values found for α, he remarks that one is derived from nine times as many data as the other, which makes it 'nine times as good' and its (probable) error nine times less. He introduces x for the error in α, and writes $\alpha = 1°30' \pm x$; the first determination from only three observations yielded $\alpha = 1°40'$ so the error therein was then $10' \pm x$. Mayer's supposition that the error behaves inversely proportional to the number of observations gives him the equality of ratios

---

[35] *"Der Vortheil aber bestehet darinn, daß durch die obige Absonderung in drey Classen die Unterschiede unter den dreyen Summen so groß geworden, als es möglich war. Denn je größer diese Unterschiede sind, je richtiger lassen sich die unbekannten Größen von α, β und θ daraus finden"* (Mayer 1750a, p. 154).

$\pm x : \frac{1}{27} = (10 \pm x) : \frac{1}{3}$, for which he gives a solution[36] $x = 1\frac{1}{4}'$. The other solution, to wit $x = 1'$, he does not mention; but he concludes that the true value of α might differ $1'$ or $2'$ from $1°30'$. Likewise, he concludes that β must be about right and that an error of $1°$ may exist in θ. Using a bootstrap technique,[37] I established standard deviations for α, β, and θ as $2.9'$, $2.6'$, and $2°7'$, respectively. Mayer recognized that the determination of θ is not very reliable because the angle formed by the lunar equator and the ecliptic (i.e., α) is so small. I am not aware whether Mayer made any other error estimates of this kind.

9.4.3 Discussion

Stigler[38] highlights the novelty of Mayer's data handling and its influence on contemporary and later mathematicians (notably Lalande and Laplace). He stresses that Mayer's treatise is remarkable for its time, because Mayer found it useful to combine so many observations, and because he attempted a quantitative error estimate. Mayer was too optimistic, in our modern view, when he supposed that the error in his determination of α behaved inversely proportional to the number of observations; yet the exploitation of the fact that a relation between the two exists at all was an important step in the theory of errors.[39]

Stigler's investigation into the intellectual climate in which the method of least squares was conceived leads him to a comparison of this work of Mayer's with slightly earlier work of Euler's on the perturbations of Jupiter and Saturn, and with later work of Boscovich and Laplace. All these people were involved in the fitting of model parameters to observations, and they did so in more or less innovative ways. Stigler concludes that Euler, as a mathematician, was wary of the accumulation of *maximum* error when observations are combined, whereas Mayer as a practising astronomer was aware (more than Euler) of the cancelling properties of random error. But, as Stigler stresses, Mayer did not go so far that he applied the property of cancellation in all cases; he allowed it only when similar data were taken under similar circumstances (i.e., same observer, same instrument, etc.; however, our remarks in Sect. 9.3 seem to contradict this somewhat). Stigler considered Mayer's division into three disjoint classes as a division among different observational circumstances, reflected in the coefficients, that Mayer preferred to keep separate. Later, Laplace would go further than Mayer by treating the

---

[36] The solution was erroneously printed as $1\frac{1}{4}''$ (Mayer 1750a, p. 155).

[37] Press et al. (1995, pp. 291–292).

[38] Here I discuss Stigler (1986, Chap. 1). Other commentaries on Mayer's *Umwälzung* tract are in Forbes (1980, pp. 48–52), Wolf (1890–1892, II, pp. 506–509), Lalande (1764, Vol. II, pp. 1234–1243), Farebrother (1998, pp. 11–15), among other places.

[39] Farebrother (1998, p. 15); also see Sheynin (1971).

set of observations as a whole. Laplace devised a general method of combining observations, which, when applied to the Manilius data of Mayer, would combine the entire corpus of 27 equations by addition or subtraction in three different ways, to arrive at a system of three equations, each depending on all 27. In Laplace's method each of the three combined equations depended on all of the original equations instead of on a subset. From Euler, via Mayer, to Laplace, Stigler signals a steady increase of the willingness to let random observational errors cancel each other.

On the other hand, Stigler signals a lack of generality in Mayer's procedure. Mayer obtained good results because of his design of the experiment and because of the geometry of the problem (and, I would add, also because of his skill as an observer). One particular circumstance in Mayer's formulation of the libration problem was the appearance of both $\sin(g-k)$ and $\cos(g-k)$ as coefficients in the equations. Upon putting those with extreme values for the former in the first and second classes, those with extreme values for the latter are necessarily left for the third class. This is a particularly helpful relation between two of the three coefficients in the equation. It is not at all obvious how Mayer's procedure would generalize in the case of more unknowns or when the relations between the coefficients are less favourable. The special circumstances of the current application rendered the criterion for class allocation quite obvious. But how would he apply this procedure to the problem of adjusting two dozen lunar equations to over a hundred observations? As the number of unrelated coefficients grows, the allocation of the observations over just as many classes gets increasingly arbitrary. At the least, it would be required to investigate the effect of the chosen allocation on the fitted parameters.

To understand why the division in classes is so effective in maximizing the differences between the coefficients of the three summed equations, we return to equation (9.2). It contains the three unknown quantities α, β, θ, which need to be determined. It also contains three known quantities g, h, k, which ultimately depend on the time of observation. These known quantities yield the numerical coefficients of the equations, in the form of $90° - h$, $\sin(g-k)$, and $\cos(g-k)$. It is immediately apparent that the last two of this triple are related – we will now see that they also have a relation with the first.

In fact, Mayer's choice to fill classes I and II with those equations that have extreme values of $\sin(g-k)$ leaves equations with extreme values for $\cos(g-k)$ to constitute class III. These class III values all happen to have the same sign, for reasons that we will explore further down. If they had had mixed signs, the equations in the third class would not sum to an equation with a large last coefficient. The situation is depicted in Fig. 9.2.

But surprisingly, a fresh look on Fig. 9.1 will show that the arc $g - k = NB$ also governs the arc $MB = 90° - h$. In fact, the spherical triangle APM has sides AP and PM of fixed measure, so that the side AM and its complement MB depend on the angle at A. It can be seen that MB is least when the triangle collapses with P in between A and M; it is largest when A

is between P and M. Thus, we see that the three numerical coefficients are interdependent.

Why do all class III equations have a positive coefficient in their last term? Stigler points out that this property is to an extent responsible for Mayer's success, but I disagree with Stigler when he suggests (pp. 22–23) that they turn out to be positive because of Mayer's choice of crater. In contrast, I will argue that they turn out positive because of the dates at which Mayer observed. In passing, however, we note that equations in class III with negative values for these terms could have been easily handled if Mayer had been willing to *subtract* them from the others in their class, instead of adding them.

Mayer's choice of the crater Manilius was governed by the necessity to select a distinctive feature close to the centre of the visible lunar disc, otherwise he would be unable to make accurate measurements of its position. Thus, from the vantage point of the moon's centre, the direction of the feature should not differ more than, say, 20° from the direction of the earth. As a consequence, the ecliptic longitude g of the feature is predominantly dependent on the ecliptic longitude of the earth, as seen from the moon; or conversely, on the longitude of the moon as seen from the earth. This quantity is periodic with period a (tropical) month. In contrast, the longitude of the ascending node changes approximately 20° during the full year of Mayer's Manilius observations, which makes it only very slowly changing with respect to g. Consequently, we see that $g - k$ completes a circuit of the ecliptic in approximately one month.

Now classes I and II have devoured those observations where $g - k$ is near 90° or 270°, leaving for class III the cases where $g - k$ is closer to 0° or 180°. When we realize that $g - k$ is approximately the arc Ωb in Fig. 9.1, it follows that class III holds those cases where the earth–moon direction is more or less along the axis $\Omega C \mho$, and the sign of $\cos(g - k)$ is positive or negative as an earth-bound observer views the moon in the direction of the descending or ascending node, respectively. Therefore, the reason why the coefficients $\cos(g - k)$ in the class III observations are negative (giving a positive last term in the equations, taking note of the minus sign in equation (9.2)) is that Mayer observed them on dates when the moon just happened to be nearer to its ascending node.

It would be illustrative to know whether Mayer planned these observing days in advance, or whether he selected a convenient set of observations a posteriori. Mayer mentioned ten extra observations of Manilius in his text, which he did not include in his working data set for various reasons: either they seemed to be less accurate, or they were inappropriate for the determination of the orientation of the lunar axis. I calculated the value of $g - k$ for those observations, whereafter it appeared that one or two might have ended up in class III with the wrong sign. This could have been the reason why Mayer rejected them.

The scatter of the values of $g - k$ (Fig. 9.2) suggests that Mayer might have put some planning in his observation schedule. However, a large proportion

Fig. 9.2 The values of $g - k$ plotted along the unit circle. The plot symbols are: *open triangle* for Mayer's class I, *open square* for class II, *open diamond* for class III, and $\star$ for the rejected data

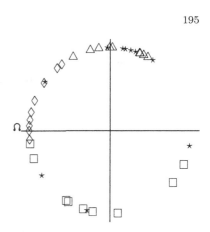

of the observations was made in July 1748, during a lunation that ended in a solar eclipse, whereafter the next lunation incidentally offered a lunar eclipse. Mayer gave ample evidence that these eclipses had his full attention, to squeeze every possible bit of information from them.[40] That is why he was making more than casual observations of the sun and moon around that time. And it just so happened that the solar eclipse occurred near the descending node, implying that most of his Manilius observations before and after the event were nearer the ascending node, when the crater was sunlit. Therefore, the values of $g - k$ populate predominantly the left half of Fig. 9.2, and Mayer had to wait half a year before he could make the rightmost observations in the figure, when the sun illuminated Manilius near the descending node. The unequal spread suggests seized opportunities rather than advance planning.

9.5 Euler's Lunar Tables of 1746

Did Mayer apply the method of averages to correct the coefficients of the lunar motion tables? It is unlikely that he did so after the advent of his spreadsheet tool during 1753. The development of the spreadsheets begins in Cod. $\mu_{41}^{\sharp}$, a manuscript that I characterized in Sect. 8.2 as extremely relevant to the various researches surrounding the publication of the *kil* tables. If Mayer had applied the method of averages *before* the spreadsheet tool, this manuscript would be the most likely place to find its traces. A search in Cod. $\mu_{41}^{\sharp}$ for systems of linear equations makes one halt at folios 273r–286v, which form the object of study of the present section. They make up a complete quire, which I have already briefly alluded to on p. 151.

---

[40] Mayer reported his results in the same (and, unfortunately, only) volume of the *Kosmographische Nachrichten*; see also the charts that Mayer and Lowitz drew in preparation of the events, probably working together, Mayer (1748) and Lowitz (1747).

The first few folios of the quire are clearly connected to Mayer's libration research: they contain calculations related to lunar craters (Manilius, Dionysius, and Menelaos) using observations taken in 1748, which Mayer included in the *Umwalzung* tract that we studied above. The ensuing folios bear 37 lunar position calculations that employ the lunar tables of Euler, *Tabulae astronomicae solis et lunae*.[41] Mayer had prepared his own manuscript copy of Euler's tables[42]. Although Mayer copied the equAtions out of Euler's tables, he adjusted the mean motions from the Julian calendar and mean time on the Berlin meridian, to the Gregorian calendar and mean time on the Paris meridian. In some of his articles for *Kosmographische Nachrichten*, he referred to these tables as the best available at the time.[43]

Clearly, this quire is out-of-sequence from a chronological point of view. Although the surrounding folios contain lunar position calculations based on the 1753 kil tables, this one was written perhaps four years earlier, when Euler's lunar tables were still the best available.

The 37 position calculations serve to compare the tables to observations of lunar meridian passages.[44] For each observation, Mayer deduces a linear equation in variables t, u, v, w, x, y, and z. For example, for the observation of twelfth February 1739 he derives:

$$1000t - 58u + 920z - 700v + 847w + 62x - 962y - 328000 = 0.$$

The constant term in all these equations is 1,000 times the difference of computed and observed lunar longitude (here, 328″). All equations have a term 1,000t, where t presumably represents a number of arc-seconds by which the epoch longitude of the moon should be adjusted. The variables t through z stand for changes to the coefficients of the tables. The coefficients of $v \dots z$ in the linear equations are the sines of the arguments of those tables, multiplied by 1,000, while t and u express adjustments to the mean motion of the moon's longitude and apogee, respectively.

Here we have Mayer deriving 37 equations in 7 unknowns, clearly doing so in the same period when he worked out the orientation of the lunar axis with the method of averages. The parallel forces itself upon us. In this case, a solution (in an approximate sense) of the overdetermined system would provide a correction of Euler's lunar tables. Interestingly, in Mayer's manuscript copy of those tables, I found the following notes written at a later instance next to the table headings: mean motion table, $t = -40″$ and $u = -12'$; table I, $z = +6'40″$; table II, $v = +2'10″$; table III, $w = -57″$. That looks like

---

[41] Euler (1746).

[42] It is now in Cod. $\mu_{14}^{\sharp}$, fol. 1–13.

[43] See for instance Mayer (1750b, §12).

[44] These observations are not related to the crater measurements above. Mayer copied most of the observations out of the *Mémoires* of the Paris Academy for the year 1739. In Cod. μ_{12} are his extracts for two-thirds of the data; I was unable to trace 12 observations of 1742 there.

a partial solution of the overdetermined system. Unfortunately, I have not been able to locate the papers where Mayer calculated this partial solution. Numerical experiments on a computer showed that a least-squares solution is unstable and liable to drastic changes when one or more observations are discarded.

With so little evidence we can only speculate. We are not sure if Mayer corrected Euler's tables before or after he successfully applied the method of averages in his libration research, although the time interval between the two events is unlikely to be more than about a year. Around that time, Mayer remarked that he had made several adjustments to Euler's lunar tables, which supports this interpretation.[45] I hypothesize that Mayer endeavoured to apply the same method of averages, but that he immediately became aware of its limitations. With more than just a few variables, it is no longer a trivial matter to decide on the distribution of equations over classes. This limitation seems to be less prevalent in Laplace's generalization of the method of averages.

Mayer's interest in Euler's tables had disappeared before the end of 1750 (see page 30). Apparently, his attempts at a systematic improvement of the Euler tables had a longer lasting value to him. Mayer inspected his earlier work in connection with fresh position calculations made during 1753, after the publication of the kil tables, therefore even after the development of the spreadsheet technique. The dislocation of these older folios shows us that the method of averages was still on his mind when he had already invented the spreadsheet tool.

Interestingly, similar sets of linear equations as were presented above, in up to eight unknowns, figure in the final chapters of Euler's treatise on the great inequAlity of Jupiter and Saturn.[46] There, Euler attempts to adjust his equAtions of the Saturn orbit to observations, and the linear equations play exactly the same role as in Mayer's work to improve on Euler's tables. Moreover, both mathematicians apply themselves to the same kind of ad-hoc strategies to get an impression of the magnitudes of the unknowns; they concentrate on those with the largest coefficients first and neglect those with the smallest coefficients, altogether in a much more haphazard way than in Mayer's later method of averages. After having fixed some of the unknowns, Euler finally tabulates what happens to the differences between observed and computed positions of Saturn when he assumes values for some of the remaining unknowns. This makes one think of Mayer's spreadsheets, and it may indeed well be that Mayer obtained the basic idea from this influential treatise of Euler.

---

[45] '... but that I made some improvements [to Euler's tables] guided by many observations...' ("... daß ich aber aus Anleitung vieler Beobachtungen einige Verbesserungen gemacht habe...") (Mayer 1750b, §12).

[46] Euler (1749a), pp. 123–141 of the reprint edition.

Chapter 10
Concluding Observations

10.1 General Conclusion

Theoria Lunae, the published treatise of Mayer's lunar theory, leaves a number of features of that theory unexplained. The most important of the unexplained features are a justification of the multistep procedure and an explanation of the process of adjusting its coefficients to observations. Mayer considered each of these features at least as important as the theory itself.

Through an investigation of the manuscripts that Mayer left behind, important new insights into this matter have here been obtained. The manuscripts show that Mayer had adapted the multistep procedure from the structure of Newton's *NTM*. This completely new insight stretches the influence of Newton's 1702 lunar theory considerably. Also, as Mayer had drawn significant attention with his lunar tables while his theory was still in a problematic state at best, we see that the role of the calculus is less preponderant than has hitherto been thought.

Mayer's awareness of the need to adjust his tables to observed positions of the moon put him in the uncomfortable, yet interesting, position that he had a large body of data but no method to deal with it. Eventually, Mayer developed an intriguing tool to help him in this respect, a tool that has a certain similarity to the modern electronic spreadsheet. It helped him make decisions about how to adjust the table coefficients. Through a study of his work we have gained more insight into the pre-history of statistical data analysis. To paraphrase Stigler's quotation on p. 3 above: working with these manuscripts has brought exhilarating senses of discovery and clearer understanding, especially when the network of interrelations between the various archive items was uncovered.

Thus, the success of Mayer's lunar tables depends on a hybrid mix of a dynamical theory based on differential calculus, a reworking of Newton's prescriptions of *NTM*, and model fitting. The latter two aspects were more persistent than Mayer's treatise on the lunar theory: the multistepped format

S.A. Wepster, *Between Theory and Observations*, Sources and Studies in the History of Mathematics and Physical Sciences, DOI 10.1007/978-1-4419-1314-2_10,
© Springer Science+Business Media, LLC 2010

of the tables survived into the early years of the nineteenth century before it was abandoned, and the principle of amassing data to fit parameters is now firmly established in scientific practice.

The next pages discuss some of our findings in greater detail. The discussion is concluded with recommendations for further research. Finally, the last section of the chapter presents an updated view of the development of Tobias Mayer's lunar tables.

10.1.1 Remarks on Data Analysis

The eighteenth century witnesses a slow change in the role of observations, from a collection of observed phenomena from which one picks examples at will to a corpus that in its totality forms part of the protocol of scientific research.[1] Mayer plays a pioneering role in this development. He had a disposition for large and conflicting data sets, and he had such a strong confidence in his technique to fit an astronomical model to observations, that he did not hesitate to export the idea to other disciplines, such as a study of the temperature distribution over the earth. His endeavours to adjust the coefficients of his lunar tables went much further than Euler's, Clairaut's, and d'Alembert's.[2]

It is remarkable that Mayer successfully fitted over 20 coefficients to over 100 observations, and at the same time it is characteristic of him. It is remarkable because nothing like it had been accomplished before and because no theory of errors or statistics was available off-the-shelf. On the other hand, it is characteristic of him, because the theme of fitting models to data is a regular one in Mayer's work. Both aspects taken together make Mayer an interesting figure in the early history of statistics. The making of his lunar tables provides a rewarding case study, not merely because the tables had important implications for practical astronomy and navigation, but most of all because of the complexity of the task of model fitting that he completed successfully.

We found that the quality of his fitted tables matches the quality of his data. This means that Mayer could only have achieved a significantly better result if he could have taken more and/or better data into account. However, the spreadsheet technique seems to have been more effective in reducing the outliers than the errors in the centre of the distribution.

Yet, his spreadsheet tool formed part of a technique, not of a method. Its application was confined to the individual of Tobias Mayer performing one

---

[1] Pannekoek (1951, p. 279).

[2] Euler used 13 lunar eclipses to adjust the moon's epoch position, eccentricity, and one equation (Euler 1753, Chap. XVII). Clairaut mentioned the possibility of adjusting but hardly changed anything (Clairaut 1752b, p. 83ff). D'Alembert's proposal for adjusting coefficients has been discussed on p. 147.

specific task, and Mayer did not carry out the generalization to a method that would allow someone else to fit for example a model of spatial temperature to data.

The so-called method of averages, in the form in which Mayer used it, was not general enough to attack the fitting of the lunar motion coefficients. Mayer made an attempt to apply it in that context, but he soon gave up. It cannot truly be regarded as a method before Laplace generalized it. Nevertheless, in the generalized form it became known as 'Mayer's method'.

10.1.2 Remarks on the Use of Data

On some occasions in the preceding chapters, we have reflected upon the re-lation between theory and observations. Even when a specific perturbation of the lunar motion has been deduced from the theory, it is still no trivial matter to verify its magnitude, or even to discern its actual existence, from real-life measurements affected by observation errors of comparable size. It is no won-der that astronomers realized relatively early that they needed quantitative tools to deal wih those errors in order to make the best possible use of all avail-able observations. The middle of the eighteenth Century was an active period in that respect and Mayer contributed actively. It takes a variety of skills to produce good lunar tables: an able observer, skilful reduction of observations, an understanding of the analytical theory, and a feeling for the handling of data and models. These skills were apparently united in Mayer's person.

Regarding the latter of these skills, he was careful to select observations under favourable circumstances, and he designed procedures and instruments with an eye on the control of errors. He was ready to discard outliers, although not always: the spreadsheet technique weighed the outliers perhaps too heav-ily. Mayer recognized the fact that random errors tend to cancel one another. Sometimes (e.g., the investigations of the latitude of Nuremberg) he aver-aged over results obtained from not equally trustworthy data sets without weighing them.

We have seen that Mayer was able to improve his tables using his spread-sheet technique almost to the utmost attainable quality: the residual errors show to us a standard deviation only little larger than that of his data set, and also very near the standard deviation that would result after a least-squares fit. This is remarkable because he seems not to have possessed a method. Yet he wasted much effort in trying to adhere to a level of exactness that was not warranted by the quality of data, as e.g., working the sheets often to $\frac{1}{2}''$.

Apparently, as was described on p. 180 above, in the second half of the eighteenth century an attitude towards data seems to have prevailed in which a *lack of accuracy* (such as the dispersion in the input data at Mayer's dis-posal) was accepted, by many even recognized as partly inevitable, while at the same time there was a concern for a *loss of precision* that was undue in view of the limited accuracy of the data. Further research could show whether this phenomenon was ubiquitous and what kind of pre-statistical thinking was behind it.

10.1.3 Remarks on the Theory

It must have been difficult for Mayer to write *Theoria Lunae*: his tables relied more on the adjustment to observations than on a theory. Moreover, he did not yet have a coherent theory from which they could be derived. He also had no theoretical justification of the multistepped format. Yet, a lunar theory was required to back up his claim to the Longitude Prize.

He needed more time than expected to complete *Theoria Lunae*. The reasons for the delay might have been the difficulty to link the single-stepped theoretical solution to the multistepped format of the tables, but further research would be needed to corroborate this.

In the introduction of *Theoria Lunae*, he prudently warned that agreement to observations was the real test of the tables, and that his theory aimed merely at showing that from the theoretical side, no objections against the tables could be raised.

Theoria Lunae has certain characteristics in common with the contemporary theories of Clairaut and Euler: all three expound the 'main problem' of the moon. Mayer's theory differs by the choice of independent variable, and also by the treatment of the latitude, for which his colleagues had used the inclination and the node position. Mayer repeatedly stressed that he did not compute the lunar latitude through the inclination and the position of the node.

While Euler was engaged in experiments with variation of constants techniques, Mayer's theory has nothing of this kind. On the contrary, his decision to treat the latitude equation in parallel with the longitude equation can be regarded as a move in the opposite direction. The independent variable that Mayer employs has a tinge of variation of constants, yet there are no signs that Mayer thought of it in that context. We are led to the conclusion that Mayer did not employ the fundamental variation of constants technique that proved to be so successful later in the century.

Another remarkable feature of Mayer's theory is that he managed to control the multitude of terms in the trigonometric series without getting bogged down in a notational nightmare. The influence of Clairaut's lunar theory on Mayer is not clear and may be elucidated through further research.

Mayer's theory may be criticized because it is not completely self-supporting: the determination of constants depends heavily on observationally obtained results, much more so than, e.g., d'Alembert thought appropriate: what the theory is supposed to provide is to some extent put into it. Mayer does not explain this in his text. It is not unlikely that the agreement between his theoretical solution of the differential equations and his lunar tables is a result of precisely this dependence.

10.1.4 Remarks on the Multistepped Format

Mayer developed the multistepped format of his tables during 1752. The format grew out of a reformulation, with the help of trigonometric functions, of Newton's *NTM* lunar theory, as embodied in Lemonnier's tables. Euler's development of a theory of trigonometric functions had been most instrumental, and the resulting tables were considerably easier to work with than Lemonnier's. But when the multistepped format is considered in the light of solutions of differential equations by trigonometric series, then it turns up quite unexpectedly, because such solutions are naturally single-stepped. In a sense, Mayer's format takes a transitional position between the older models of lunar motion with kinematic elements and the newer, dynamical, theories that were emerging at the time.

Mayer's change of strategy seems to have been made for pragmatic rather than theoretical reasons. He made this move when his theory provided a disappointing accuracy, while looking at the work of other colleagues did not provide adequate inspiration. His first version of the new design provided an accuracy in the predicted positions of the moon that was four times as good as his earlier theory and slightly more accurate than Lemonnier's tables. The reason for the surprising increase in accuracy after adoption of the new scheme should be the subject of further research.

In the preface to his first printed tables of 1753, Mayer offered remarkably little in support of an alleged dynamical theory of the moon's motion. Mayer carefully avoided to assert that he indeed had such a theory, whereas the kinematics of Newton's *NTM* clearly shows through. Mayer's credits to Euler must be understood not in the context of a dynamical lunar theory, but rather in connection with the changing perception of trigonometry which the latter had brought about.

The theory that Mayer later worked out in *Theoria Lunae* yielded a single-stepped solution in the first place, which he transformed into the multistep one that was also given in that treatise. In order to perform the transformation, he needed an approximative procedure to reduce sines-of-sines into sums of sines. Such a procedure, which can also be found in d'Alembert (1754–1756), was apparently not widely known, perhaps because the particular type of problem that it solves does not occur very often.

Mayer's stated motivation for his multistepped format was that it reduced the number of significant terms. In Chap. 7, this was proved to be false at least for the later multistepped format with evection separated from the equation of centre. It has not yet been investigated whether his argument also fails for the original multistepped format. But given that Bradley had already questioned the change of the format between the submitted manuscript tables and the published earlier ones, Mayer would have been unwise to change the format again or even to give up on it. It is not unlikely that he regarded the multistepped scheme as a hallmark of, if not a major innovation in, his lunar tables. Even if he had discovered the futility of it during his work on *Theoria Lunae*, he is well excused to maintain it.

10.2 Further Research

A number of items have turned up that deserve further research. Here they will be briefly summarized.

It would be illuminating to compare the standard deviations of Mayer's lunar tables (Chap. 6) to those of the various theories of Euler, Clairaut, and d'Alembert. Computing power is nowadays easily available to almost every researcher. This makes it feasible to investigate and compare the accuracy of various historical lunar theories. Examples of such investigations can be found in the current work and (among others) in Kollerstrom (2000) and Thoren (1974). But since all these investigations are geared towards their own specific goals, their results are hard to compare. This makes it nearly impossible to tell how good or bad any specific lunar theory performed. A comprehensive study, consistently carried out, of all the past lunar theories should be undertaken in order to provide an authoritative reference.

We have signalled that a discrepancy occurred between the stated accuracy of results and their inherent precision. This discrepancy is found for instance in Mayer's spreadsheet contents and also in the accuracy with which mariners were expected to work their lunar distances (see p. 180). It would be interesting to know how common this phenomenon was and how and when the practice changed into the modern scientific one. This would further our understanding of scientific attitudes in the era of emerging statistical awareness.

Mayer repeatedly averred that his multistepped format made many terms of his lunar theory negligible. His claim is not substantiated by our investigations in Sect. 7.2. Therefore it is tempting to conclude that the accuracy of his tables was brought about not so much by the multistepped format, but rather by some other factor, such as a particularly happy choice of coefficient values. But before such a conclusion may be drawn, we have to consider if his claim also fails for his original multistepped format, where evection and equation of centre appeared conjointly. After all, Mayer's success came with his original implementation of the multistepped format, while we debunked his claim by the final format. Was his claim perhaps true for the original form of the multistepped procedure? If so, for what reason does it fail with the later format?

The most important change to the multistepped format was the relocation of evection, which has considerable implications for the coefficients of many equAtions. Mayer calculated the effect on these coefficients on foll. 52–53 of Cod. μ_7. These calculations have still to be studied in order to establish the technique that he used.

In *Theoria Lunae*, Mayer compared the theoretically derived coefficients of the equAtions to the coefficients of his lunar tables. The differences between these two sets were generally smaller than the differences between any of these and any of the lunar theories of Euler, Clairaut, and d'Alembert, as Mayer remarked in the preface of his theory. Thus, Mayer concluded that his theory and tables matched particularly well, so that the theory offered at least nothing against his tables. However, we now know that he made use of his table coefficients for the computation of the theoretical coefficients; therefore the question arises whether this match is a valid argument in support of the tables.

The many lunar theories that have been developed over time, are interconnected in many, sometimes obscure, ways. We have seen that Mayer's theory was influenced by work of Euler as well as by Newton and Horrocks. Less clear is the influence of d'Alembert and Clairaut. Clairaut's lunar theory had possibly two direct influences on Mayer: the method of computation of the coefficients, as outlined in Sect. 5.4.9, and the inversion of series. Both were treated to some extent in Clairaut's theory. It is conceivable that Mayer's choice of independent variable was also inspired by Clairaut's lunar theory. There may have been other influences as well. An investigation of Mayer's planetary theories (especially of Mars and Jupiter) and their relation to other theories is also highly desired.

Turning now to the spreadsheets, we note a puzzling anachronism that is begging for an explanation. The sheets in Cod. $\mu_{41}^{\sharp}$, which relate to the kil tables of 1753, look as if they are more primitive and rudimentary than the others, which would indicate that they are the earliest spreadsheets. But some sheets in Cod. $\mu_{28}^{\sharp}$, although they look more evolved, are apparently related to older put and zwin table versions. In case that the former are indeed the oldest spreadsheets, it must still be explained why Mayer returned to the older table versions in order to fit them with his new technique; but if they are not the oldest, then their primitive appearance must be explained.

So far as the spreadsheets have been investigated, they seem to be used to fit the longitude equAtions only, including the epochs and mean motions. I found no spreadsheets that address latitude or parallax. Analysis of the remaining sheets that have not yet been investigated might reveal whether Mayer ever fitted the latitude and parallax coefficients with his spreadsheet technique.

A completely different question arises from Mayer's early work on lunar theory. In 1751, he had obtained a formula to use the Saros as a tool in predicting lunar positions, as was proposed by Edmond Halley.[3] It is currently

---

[3] Mayer wrote about this to Euler on 4 July 1751 (Forbes 1971a, p. 34).

not known how Mayer had obtained that formula, nor why he abandoned it. Although the use of the Saros in this way seems to have had little impact on his work, a further investigation of this matter might illuminate his approach of the lunar motion problem at that time.

Mayer's sincere interest in collecting data and making the best possible use of them stretched to a number of other areas. His proposal for a model of world temperature ditribution has already been mentioned. A further opportunity to investigate his data handling ambitions exists in his collection of magentic data and associated study of the geomagnetic field. Such an investigation should start with the third volume of Forbes (1972) and the manuscripts mentioned therein.

10.3 The Development of Mayer's Lunar Tables: A Summary

Since the current study has turned up quite some new insights into Mayer's work on lunar motion, it seems fitting to end here with a review of the main events in that work. We start with a general outline and elaborate on specific topics further on.

Tobias Mayer's lunar tables of 1753 are of complex origin. They incorporated elements of Newton's *Theory of the Moon's Motion* with certainty, elements of a dynamical theory in all likelihood, and they had their coefficients adjusted with the help of observations. They originated certainly not merely from Euler's lunar theory, fitted to observations: neither was the content of Euler's theory known to Mayer at the time of publication of the tables, nor would it provide an inducement for the multistepped format of the tables.

Mayer's tables facilitated positions of the moon with a standard deviation of about 45″, so that determination of longitude within the limits set by the Longitude Act became a realistic possibility. Urged by Euler, Michaelis, and other colleagues, Mayer devised a method to find longitude at sea, consisting of improved lunar tables, a description of a repeating circle (to measure the required angles more accurately), and prescriptions for the computation that had to be performed. Late in 1754, these were bundled and sent as a package to the Board of Longitude, whereupon Bradley requested an explication of the principles upon which the tables were based.

Mayer worked on the requested theory confidently during January and February of 1755, but unexpected complications delayed its completion until November. It is probable that difficulties to bring the theory into accord with the format of the tables were responsible for the delay.

Because the Seven Years' War intervened, the Board did not reach a decision before Mayer's death. On 9 February 1765, the Board proposed to British Parliament that rewards should be granted to the clockmaker Harrison and to Mayer's widow. Parliament decided that Harrison should be rewarded

£10,000 for his watch (and double the amount when he demonstrated how to produce many more), the widow £3,000, and that additionally Euler should receive £300 for his contributions to the theory of the moon's motion. With hindsight, the reward to Euler seems to be not wholly appropriate, because Mayer's success depended less on a sound theory than on an accurate fit of the parameters. Mayer was certainly an admirer of the great man and he learnt many techniques from Euler's writings. A significant source of the confusion concerning *Theoria Lunae* was Mayer himself, who never skipped an occasion to express his debts to Euler.

At the same meeting of the Board of Longitude, the new Astronomer Royal, Nevil Maskelyne, was charged with the production of an annual Nautical Almanac, containing precomputed lunar distances based on the 'last manuscript tables' which Mayer's widow had sent in 1763. The first Nautical Almanac was produced for the year 1767. Mayer's 'last manuscript tables', eventually improved and augmented by Mason and Bürg, continued to influence the Nautical Almanac until the beginning of the nineteenth century. The tables were published in various publications such as Mayer (1770), Lalande (1764), and Hell and Pilgram (1772).

The multistepped format of the 1753 tables stemmed from the format of Newton's *Theory of the Moon's Motion*, which provided the most successful lunar tables extant during the second quarter of the eighteenth century and which enjoyed widespread distribution predominantly through Lemonnier's *Institutions Astronomiques*. Mayer's adaptation resulted in tables that were more straightforward to use than Lemonnier's. Besides, after modelling his tables in the multistep way, Mayer reached a fourfold increase in accuracy over his former tables, without an increase in the number of equAtions. The reasons for this success lay largely in the lucky choice of coefficients, although it may be that its rendering of the Horrocksian mechanism of the variable eccentricity and oscillating apsidal line has meant a significant contribution, too. Mayer's lunar theory in the dynamical sense was still in an unsatisfactory state at the time.

The improved tables of 1754 with which Mayer entered the quest for the Longitude Prize, as well as the last manuscript tables on which the first almanacs were based, incorporated a streamlined multistep format that was easier to work with than the format of the 1753 tables. This change broke the link with the Horrocksian mechanism. At that time, Mayer could well have done away with the multistepped format, which complicated the relation between the theory and the tables. However, Mayer mistakenly averred that it reduced the size of some equAtions to below the limit where they could be neglected.

Because Mayer had continued to adjust the coefficients to observations, the standard deviation of predicted lunar positions had come down to about 30″. To adjust the coefficients, Mayer used over a hundred observations of occultations and eclipses of the seventeenth and early half of the eighteenth century; after 1755 he included perhaps a significant fraction of his own lunar observations. It was an immense operation to 'fit' more than twenty coefficients to over a hundred observations without a statistical method, being guided mainly by numerical and/or astronomical insight. To keep track of the effects of changes to the coefficients, Mayer devised a tool with a functionality similar to modern electronic spreadsheets. The organization of data into these spreadsheets helped him to quickly investigate the effect of amendments to coefficients. His knowledge of the subject matter, combined with his numerical abilities, guided him in the choice of amendments to make. It was not so much Mayer's observational skill, but rather his careful handling of observations and his numerical abilities that were responsible for the success. His perseverance to use a multitude of observational data, apart from the skilful handling of them, was responsible for the accuracy of his lunar tables. This was a characteristic feature of his work in general. Mayer was a pioneer in data analysis.

The reputedly close resemblance between the lunar theories of Euler (1753) and Mayer (1755) goes as far as the formulation of the differential equations and ends there. Mayer developed an approximative solution to the differential equations in his own way. He used techniques that were modern for the time, and his theory fits in well among the best lunar theories of his time, such as Euler's and Clairaut's. Of Euler's theory we can be sure that it had an impact on Mayer's *Theoria Lunae*; Clairaut's theory may well have had an influence too. It is remarkable that Mayer chose an independent variable that had no physical interpretation.

Mayer's lunar theory does not contain terms of a higher order than may be found in Euler's and Clairaut's theories, nor does it include physical causes absent in those theories, such as perturbations due to the earth's shape. Yet Mayer's theory comes considerably closer to the true motion of the moon than the other theories do. It is not evident what causes this success of the theory. The determination of the coefficients in the solution leaned heavily on available empirical knowledge, in the form of coefficients of lunar tables which had already been fitted to observations. This circumstance is likely to have had a positive influence on the apparent accuracy of the solution.

Moreover, Mayer maintained that his tables should be judged by comparison to observations rather than by verification of the theory. Indeed, theory alone was at that time still unable to accomplish the measure of accuracy that Mayer's tables provided.

Appendix A
Lunar Equations: Versions and Aliases

This appendix contains information on the various versions of Mayer's lunar equAtions, in what locations they appear and in what form they appear. Display A.1 lists this information and Display A.2 shows the coefficients of some of the most prominent versions.

Most entries in Display A.1 show an indication of the form in which the equAtions appear: 'tables', as ready to use tables; 'list', as a list of coefficients and arguments; and 'comparison', as a column in a tableau where different versions are put next to each other. Locations are abbreviated as EM for Euler–Mayer Correspondence,[1] NA for Nautical Almanac,[2] TL for *Theoria Lunae*,[3] and the usual manuscript abbreviations as listed in Appendix C. Dates are mostly indicative. The following remarks apply to the individual instances of lunar tables and equAtions.

Duin is Mayer's manuscript copy of Euler's lunar tables.[4] Terp, like duin, uses eccentric anomaly, which Mayer used nowhere else in lunar tables. Therefore (and because duin and terp occur in the same manuscript) they can be considered as near-contemporary. The coefficients of kreek are not known, but the characteristics that Mayer reported to De l'Isle of these lost tables are met by certain position calculations (as mentioned on p. 30); the same characteristics are not met by any other versions described here. In principle, then, the kreek coefficients could be reconstructed from those position calculations.

Zand, diep, and gors are very nearly identical; Mayer reported zand in a letter to Euler written on 6 January 1752, therefore we have an upper bound for its dating. According to the table headings in Cod. $\mu_{15}^{\sharp}$ (see Fig. 6.4), gors was a fit of geul to observations. Therefore geul most likely dates from (late) 1751.

In reply to Mayer's letter containing zand, Euler sent Clairaut's coefficients, which Mayer included in Cod. $\mu_{15}^{\sharp}$. This helps to date griend

---

[1] Forbes (1971a).

[2] RGO-NA (1767,...).

[3] Mayer (1767).

[4] i.e., The tables in Euler (1746).

and grond, and additionally it puts a lower bound on the date of geer further on in the same manuscript, the version that clearly started the multistepped development stage as described in Sect. 6.6.

The development from geer via gat, put, and plaat into zwin, pas, and finally kil can be traced out by analysis of the position calculations in Cod. $\mu_{41}^{\sharp}$ in combination with a double folio of spreadsheets in Cod. $\mu_{28}^{\sharp}$. Zwin was reported in a letter to Euler on 7 January 1753. Only slight changes separate zwin from the kil tables that were published only a few months later in Mayer (1753b). Mayer warned his readers that the example computation that went with the kil edition was based on an older version, and indeed it still employed the zwin tables.

The post-kil tables and lists can be divided into two groups: the ones that have, like kil, the equAtion of centre and evection together in the middle step, and the newer ones with evection relocated in the first step. Even when explicit information on the location of evection is lacking, it is often possible to determine to which group a version belongs by looking at the coefficients for the equAtions with arguments 2ω or $2(\omega - p)$, if present, because their magnitudes are very much affected by the relocation.

The tables in the first group include val, vlije, vlakte, and mui. All these have the equAtion of argument $2\delta - p$ moved to the end of the procedure. Vlakte and vlije form a pair that complement each other, not only by their equAtions but also by their physical location as two pamphlets stuck together. Vlakte was in use in March 1754, when Mayer answered to Euler the question about an averred Chinese eclipse reported by Gaubil. Mayer's computations regarding that request are extant in one of the last folios of Cod. $\mu_{41}^{\sharp}$.[5]

Apparently mui was used with the computations in Cod. μ_5. Early in that manuscript, there are dates 1754 March and April; later Mayer added a calculation for 2 Oct. 1754. Thus mui dates from between March and October of that year.

Computations necessary to move evection away from the equAtion of centre are on foll. 52–53 of Cod. μ_7. These start out with a variant of the vlakte and mui coefficients; they are immediately followed by a list of the new coefficients of ley, now with evection among the minor equAtions (albeit in position VII instead of V).

The version that Nevil Maskelyne repeatedly referred to as the 'first manuscript tables' were the rak tables that Mayer sent to England around the beginning of December 1754. It seems that these are no longer extant but Maskelyne extracted and published their coefficients. They are almost identical to ley and have evection as equAtion number V. Thus the relocation computations in Cod. μ_7 just mentioned were made in October or November 1754.

Quickly afterwards, in April of the following year, Mayer wrote a letter to Michaelis stating not only his reasons for the relocation of evection, but

---

[5] This episode is recounted in Forbes (1980, p. 147), the exchange between Euler and Mayer is recorded in Forbes (1971a, pp. 79, 80, 84), and Mayer's fuller account can be read in Forbes (1972, I, pp. 96–98).

also several further amendments of the coefficients, which perfectly match the transition from rak to wijd.[6]

Wantij was originally computed in Cod. $\mu_{28}^{\sharp}$; it is the multistepped form computed from the single-stepped theoretical solution of *Theoria Lunae*, waard. Wad is wantij with certain adjustments obtained from observations: these adjustments, as Mayer dutifully reported in *Theoria Lunae*,[7] consist of a revised lunar eccentricity and solar parallax on the basis of the maximum observed elliptic and parallactic equations, respectively, and a changed evection coefficient. Wad is not the result of a complete fit of all the coefficients to observations independently. *Theoria Lunae* included a comparison between wad and wijd (his latest fitted version at that moment) in order to show to what extent theory and tables agreed. (As we have seen in Chap. 5, the computation of the coefficients in the theory made such an abundant use of fitted table coefficients, that we may doubt whether the test was unbiased.)

There remains one set of tables in Cod. $\mu_{49}^{\sharp}$, veen. This one is of the kind that has evection with the minor equations, unlike the other versions contained in that manuscript. It is practically identical to rede, the 'last manuscript tables' as Maskelyne called them, which were sent by Mayer's widow to England in 1763. The rede tables were printed in Mayer (1770); the original manuscript tables but without table headings are in RGO4/125 (second quire), the crudely written headings in mixed Latin and German are in RGO4/130.[8] Maskelyne listed the coefficients of kil and rede in the Nautical Almanac for the year 1774, together with two variant coefficient lists devised by Bradley and Morris after Mayer's tables. One page was torn out of the veen quire and sent to England together with the pack of last manuscript tables, (probably conforming to instructions that Mayer gave on his death bed), because it contains a corrected secular acceleration table on one side. On the reverse side is half of veen's evection table.

Oever seems to be a pre-version of wantij. The first few pages of Cod. μ_8 show the transformation of multistepped rede into the equivalent but single-stepped schor. The latter shows up under the heading *juxta tabb. noviss.* next to sloe, which, headed by *juxta calc. theor. corr.*, looks like a revised version of the very similar waard. The list also included the coefficients of the theories of d'Alembert and Clairaut and was presumably meant to compare the various results. Finally, the coefficients of balg (in particular, the coefficients of equations $2\omega - 2p$ and 2ω) suggest that it belongs to the same family as rede, but otherwise balg is quite a-typical. It appears in Cod. μ_4 after pages of lunar theory calculations of a later date than the completion of *Theoria Lunae*.

---

[6] The letter was published in Mayer (1770, pp. 43–45); Mayer's reasons are quoted on p. 56 above. Coefficients of wijd are also included in *Theoria Lunae* (Mayer 1767, p. 52).

[7] Quoted on p. 144 above.

[8] Actually two sets of the rede tables are extant in RGO4/125; one was sent by the widow 2 years later (Forbes 1966, p. 109). There are also revised mean motion tables referred to the Greenwich meridian, which were presumably made by Maskelyne or one of his assistants.

Display A.1 Versions of lunar equations found among Mayer's papers

Alias	Date	Source	Comments
duin	1746	Cod. $\mu_{14}^{\sharp}$	Tables, copy of Euler's of 1746
terp		Cod. $\mu_{14}^{\sharp}$, f.14+	Tables
kreek	Jan. 1751	lost	Reported in correspondence
geul	1751	Cod. $\mu_{15}^{\sharp}$, f.8v	List *calculus m. ex theor.* (fig. 6.4)
gors		Cod. $\mu_{15}^{\sharp}$, f.8v	List *corr ex observ* (fig. 6.4)
diep		Cod. $\mu_{14}^{\sharp}$	List
zand	Jan. 1752	EM p.48	List, from correspondence
griend	Mar. 1752	Cod. $\mu_{15}^{\sharp}$, f.8v	List *tabb ☽ calc* (fig. 6.4)
grond	Mar. 1752	Cod. $\mu_{15}^{\sharp}$, ff.9-16v	Tables
geer	later 1752	Cod. $\mu_{15}^{\sharp}$, f.17r	*Entwurf* (display 6.2)
gat		Cod. $\mu_{15}^{\sharp}$, f.17v+	Tables, very simple
put		Cod. $\mu_{41}^{\sharp}$, f.11-30	Reconstructed from computations
plaat		Cod. $\mu_{41}^{\sharp}$, f.108	Rables, draft
zwin	Jan. 1753	EM p.62	List, from correspondence
pas		Cod. $\mu_{41}^{\sharp}$, f.150v	list
kil	spring '53	Mayer (1753b)	Tables published before 7 May 1753
val	1754	Cod. $\mu_{49}^{\sharp}$	Tables, 1st booklet, neat, almost complete
vlakte		Cod. $\mu_{49}^{\sharp}$	Tables, 3d booklet
vlije	Mar. 1754	Cod. $\mu_{49}^{\sharp}$	Tables, booklet stuck inside vlakte
mui		Cod. $\mu_{30}^{\sharp}$	Tables, only one sheet remains
ley	fall 1754	Cod. μ_{7}, f.54v	List
rak	Nov. 1754	NA 1774	List, 'first manuscript tables'
wijd	Apr. 1755	TL p.52	List, *tabulas...ultimam correctionem*
waard	mid 1755	TL p.46-47	List, single-stepped, see display 7.1
wantij	1755	Cod. $\mu_{28}^{\sharp}$,II	List, also in *Theoria Lunae*, see display 7.1
wad	1755	TL p.52	List *ex calc*, improved by observations
veen		Cod. $\mu_{49}^{\sharp}$	Tables, one p. dislocated to RGO4/125(3)
rede	after 1755	NA 1774	'Last manuscript tables'
oever	1755?	Cod. $\mu_{28}^{\sharp}$,III,50	List
schor	post-rede	Cod. μ_{8}, f.7r	List, *juxta tabb. noviss*
sloe		Cod. μ_{8}, f.7r	List, *juxta calc. theor. corr.*
balg		Cod. μ_{4}, f.58-59	List

Display A.2 The following remarks apply to Display A.2 containing the coefficients of some important versions of tables and equations.

Epochs and mean motions are given for the moon's longitude and the position of the apogee and ascending node. The epochs are stated for 1750.0, i.e., the noon preceding 1 January 1750. The mean motions are stated as increments (for longitude and apogee) or decrement (for the node) over 60 Julian years of 365.25 days each, modulo 360°. The listed kil epochs and mean motions were extracted from the published version of 1753, and have been adjusted from the Paris meridian to the Greenwich meridian by adding $+9^{\mathrm{m}}16^{\mathrm{s}}$. The rak epochs were copied from the preface of the Nautical Almanac for the year 1774. The rede epochs and mean motions were taken from Mayer (1770); their epochs agree to what is published in the Nautical Almanac just quoted. Two arcseconds were added to all longitude epochs for secular acceleration.

The annual equation of anomaly (not apogee) and node are labelled by 'An.A' and 'An.N', respectively; the annual equation of longitude is always number I. Note that vlakte has an additional equation of $1'12''$ maximum which depends on the longitude of the ascending node and which applies to the anomaly.

In the list of mui coefficients, the symbol † indicates those coefficients that have been reconstructed from position calculations. Double entries occur in this column because some of the mui tables had been corrected by a pasted-on new version; the first listed entries are the original ones.

Display A.2 Coefficients of the most prominent versions (continued on next page)

	kil		vlakte	vlije	mui	
Epoch	6ˢ 8°22'11"					long
	5ˢ20°56'50"					apog
	9ˢ10°19' 8"					node
Mean	1ˢ10°43'24"					long
Motion	9ˢ11°30'45"					apog
	2ˢ20°30'45"					node
An.A	+20'36"	An.A	+25'15"	+22'28"		sin(ς)
	−18"		−16"	−14"		sin(2ς)
			1'12"			sin(Ω)
An.N	+10'18"	An.N	?	+9'12"		sin(ς)
	−9"		?	−6"		sin(2ς)
I	+11'20"	I	+11'14"	+11'14"		sin(ς)
	−10"		− 7"	−7"		sin(2ς)
II	−54"	II	−56"	−58"	†−56"	sin(2ω + ς)
III	−1' 2"	III		−1'6"	−1' 6"	sin(2ω − ς)
IV	+1'48"	IV		+1'50"	+1'50"	sin(2ω − p + ς)
V	+1'12"	V		+57"	+57, +49"	sin(2ω − p − ς)
VI	+1'30"	VI		+1' 7"	+1' 4"	sin(2ω + p)
VII	+58"	XIV	+1'25"	+1'25"		sin(2δ − p)
VIII	+40"	VII		+30"	+14, +28"	sin(p − ς)
IX	−47"	VIII		−56"	−56"	sin(2δ − 2ω)
X	+29"	IX	+16"	?	†+23"	sin(ω − p)
	+3'24"		+3'31"		†+3'26"	sin(2ω − 2p)
		X	+7"			sin(2ω − 2s)
		Xa	−6"			sin(2ω − 3p)
			4"			sin(Ω)
XI	−6°18'26"	XI	−6°18'18"		†−6°18'18"	sin(p)
	+13' 0"		+12'53"		†+12'53"	sin(2p)
	−36"		−35"		†−34"	sin(3p)
XII	−1°20'42"	XII	−1°20'50"		†−1°20'52"	sin(2ω − p)
	+35"		+27"		†+32"	sin(4ω − 2p)
		XIIa	+5"			sin(4ω − p)
XIII	−1'55"	XIII	−1'55"			sin(ω)
	+40'21"		+40'15"			sin(2ω)
	+2"		+2"			sin(3ω)
	+17"		+14"			sin(4ω)
XIV	−6'57"	XIV	−6'51"	−6'51"		sin(2δ)

Display A.2 Coefficients of the most prominent versions (continued from previous page)

	wad	rak	wijd	veen	rede	
Epoch	6ˢ 8°22′39″				6ˢ 8°22′26″	long
	5ˢ20°55′59″				5ˢ20°55′54″	apog
	9ˢ10°18′38″				9ˢ10°19′ 8″	node
Mean					1ˢ10°44′ 9″	long
Motion					9ˢ11°30′45″	apog
					2ˢ20°30′45″	node
An.A	+22′ 6″	+25′15″		+23′12″	+23′12″	$\sin(\varsigma)$
	−15″	−16″		−6″	−6″	$\sin(2\varsigma)$
An.N		+9′13″		??	+8′50″	$\sin(\varsigma)$
		−6″		??	−2″	$\sin(2\varsigma)$
I	+11′39″	+11′14″	+11′14″	+11′18″	+11′16″	$\sin(\varsigma)$
	−10″	−4″	−4″	−4″	−4″	$\sin(2\varsigma)$
II	−58″	−56″	−56″	−53″	−54″	$\sin(2\omega + \varsigma)$
III	−1′13″	−1′ 6″	−1′ 6″	−1′ 8″	−1′ 9″	$\sin(2\omega - \varsigma)$
IV	+55″	+53″	+49″	+53″	+54″	$\sin(2\omega + p)$
V	−1°20′ 8″	−1°20′36″	−1°20′36″	−1°20′33″	−1°20′33″	$\sin(2\omega - p)$
	+38″	+38″	+26″	+35″	+35″	$\sin(4\omega - 2p)$
VI	+2′11″	+2′ 0″	+2′ 0″	+2′ 8″	+2′ 9″	$\sin(2\omega - p + \varsigma)$
VII	+44″	+39″	+47″	+47″	+49″	$\sin(2\omega - p - \varsigma)$
VIII	+40″	+28″	+28″	+34″	+34″	$\sin(p - \varsigma)$
IX	−1′26″	−56″	−51″	−56″	−56″	$\sin(2\delta - 2\omega)$
X	+8″	+16″	+16″	+16″	+16″	$\sin(\omega - p)$
	−1′ 2″	−57″	−1′ 0″	−1′ 0″	−1′ 0″	$\sin(2\omega - 2p)$
Xa			+4″			$\sin(\Omega)$
XI	−6°18′11″	−6°18′11″	−6°18′11″	−6°18′17″	−6°18′15″	$\sin(p)$
	+13′ 2″	+12′52″	+12′52″	+13′ 0″	+13′ 0″	$\sin(2p)$
	−36″	−37″	−37″	−37″	−37″	$\sin(3p)$
				+2″	+2″	$\sin(4p)$
XII	−1′55″	−1′55″	−1′55″	−1′56″	−1′55″	$\sin(\omega)$
	+35′42″	+35′47″	+35′47″	+35′44″	+35′43″	$\sin(2\omega)$
	+1″	+2″	+2″	+1.3″	+2″	$\sin(3\omega)$
	+5″	+14″	+14″	+11″	+12″	$\sin(4\omega)$
XIII	+1′15″	+1′25″	+1′26″	+1′25″	+1′23″	$\sin(2\delta - p)$
XIV	−6′51″	−6′51″	−6′51″	−6′44″	−6′43″	$\sin(2\delta)$

Appendix B
Spreadsheet Contents

This appendix lists the results of the spreadsheet analysis that was described in Sect. 8.5.3. The amendments that Mayer made to the equAtions, and that appear on the analyzed sheets, are all listed. Every amendment corresponds roughly to a column in a sheet, but some intermediate columns (listing argument values, equAtion values, etc.) have been skipped here. All values are in arc-seconds unless stated otherwise.

Symbols indicate whenever Mayer applied an amendment to eclipses only (☉ ☽), to occultations only (⋆), when he revised an earlier one in the same sheet ($\mathcal{R}$), and when he crossed out a column (≀). A horizontal line indicates that Mayer rejected all amendments and started anew from the initial errors. The notation $+\frac{1}{3}(Y - 1700)$ signifies that Mayer made an amendment to the lunar mean motion of nominally $\frac{1}{3}''$ per year (or $20''$ in 60 years) from the epoch 1700; one of the obscurities of Mayer's approach is his seemingly random choice of different epochs.

The error values that appear in the spreadsheets are here summarized, column-wise, by box plots. These show the quartiles of the error distribution: the minimum and maximum values are at the extremes of the plot, the median is located at the vertical bar inside the box, while the box itself contains the centre 50% of values. The width of the box is also known as the inter-quartile range (IQR).

The estimated standard deviation σ has been indicated next to the box plots, for the initial and last columns of the sheets and at the place where it reaches its minimum value, excluding partially filled columns. If the error values followed the normal distribution, then $\sigma \approx 1.35$ IQR.[1]

Since the number of observations N is of so much influence on the dimensions of the plot and the value of σ, this number has been indicated wherever necessary; a ⋆ or ☉ ☽ imply a smaller value for N (but the smaller number has not been indicated explicitly).

---

[1] Rice (1995, p. 372).

The final display B.1 on the last page of the appendix attempts to give an overview of all amendments in all analyzed spreadsheets. The display has a column for every sheet, and it lists the arguments of the equations one per line. The sheets are arranged in their hypothesized approximate chronological order. For every amendment to a specific equAtion on a specific sheet, there is an entry in the corresponding row and column of the display. If the equAtion is amended more than once in the sheet, then there are multiple entries. Whenever Mayer restarts a sheet from his initial situation, the display shows a vertical line.

Some equAtions combine several terms (such as the annual equAtion, which incorporates the terms $c_1 \sin(\varsigma)$ and $c_2 \sin(2\varsigma)$); these have arguments ς, $\omega - p$, $2\omega - p$, p, and ω. The dominant terms in each of these composed equAtions are the terms with arguments ς, $2\omega - 2p$, $2\omega - p$, p, and 2ω, respectively.

Most amendments are given as a fractional change of the original coefficient. For instance, consider equAtion I, the annual equAtion of about $+11'14'' \sin(\varsigma) - 10'' \sin(2\varsigma)$. When an amendment of $-1/120$ is listed for this equAtion, then this means that the first coefficient ($11'14''$) is decreased by about $5\frac{1}{2}''$; the change in the second coefficient is negligible. The notation '$+3''[2\varsigma]$' is an abbreviation for a change of the second coefficient to $-7''$.

Amendments are here represented in their accumulated form, i.e., successive amendments to the same equAtion have been added together. Space permitting these have been maintained in their form as originally specified on the spreadsheets, e.g., $1/5 + 1/25$ instead of $6/25$, since the first form suggests more clearly how Mayer arrived at the amendment.

In Mayer's sheets, there are numerous amendments which we may here represent by the form $c \cos p$, where c is a constant and p the lunar anomaly (mean or intermediate). See sheet 8, third amendment (below) for an example. In the final display, I have reduced those amendments to changes in the lunar apogee position. The relation between $c \cos p$ and a shift of apogee is explained as follows. Note that the argument p is lunar apogee subtracted from lunar longitude, so a shift in apogee of magnitude $+\Delta p$ changes that argument into $p - \Delta p$. Denoting the equAtion of centre before and after the amendment by $\mathcal{C} = \mathcal{C}(p)$ and $\mathcal{C}' = \mathcal{C}(p - \Delta p)$, respectively, we have:

$$
\begin{aligned}
\mathcal{C}' &= \alpha_1 \sin(p - \Delta p) + \alpha_2 \sin 2(p - \Delta p) + \alpha_3 \sin 3(p - \Delta p) \\
&\approx \alpha_1 \sin p + \alpha_2 \sin 2p + \alpha_3 \sin 3p \\
&\quad - (\alpha_1 \Delta p \cos p + 2\alpha_2 \Delta p \cos 2p + 3\alpha_3 \Delta p \cos 3p) \\
&= \mathcal{C} - \Delta p(\alpha_1 \cos p + 2\alpha_2 \cos 2p + 3\alpha_3 \cos 3p).
\end{aligned}
\tag{B.1}
$$

Since $\alpha_1 \approx 6\frac{1}{3}°$, $\alpha_2 \approx 13'$, and $\alpha_3 \approx \frac{1}{2}'$, we see that a change of Δp in lunar apogee position gives an approximate change of $-\Delta p\, \alpha_1 \cos p$ in equAtion of centre; higher order terms are negligible.

The rationale for expressing such a cosine amendment here as a shift of apogee is that Mayer *intended* to apply the latter, but that he needed the former to compute its numerical effect in the spreadsheet. It is his intention that I wish to express in the display.

It is illustrative to compare Mayer's results to a modern least-squares fit of the equations to the same data that he started out with, i.e., the combination of sheets 1 (with his initial set of solar eclipses and occultations) and 6 (idem, for lunar eclipses). The first diagram below shows the box plots both before and after the fit.

Box plots of errors before and after modern least-squares fit on Mayer's data of sheets 1 and 6 combined. $N = 73$.

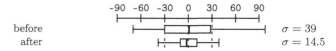

Cod. $\mu_{28}^{\sharp}$, Tab.I. $N = 68$, of which 26 are eclipses and 42 occultations.

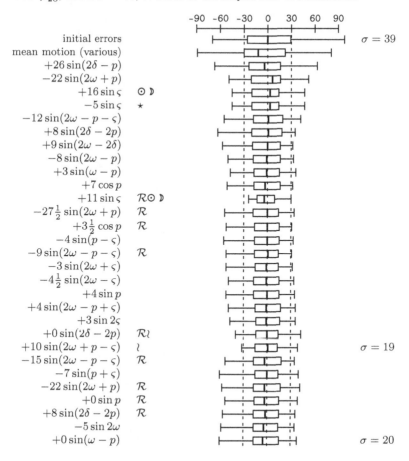

Cod. $\mu_{41}^{\sharp}$, fol. 44. $N = 13$, all are occultations of Aldebaran.

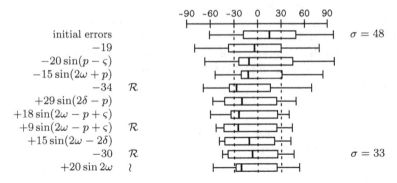

initial errors — $\sigma = 48$
-19
$-20\sin(p - \varsigma)$
$-15\sin(2\omega + p)$
-34 $\mathcal{R}$
$+29\sin(2\delta - p)$
$+18\sin(2\omega - p + \varsigma)$
$+9\sin(2\omega - p + \varsigma)$ $\mathcal{R}$
$+15\sin(2\omega - 2\delta)$
-30 $\mathcal{R}$ — $\sigma = 33$
$+20\sin 2\omega$ ι

Cod. $\mu_{41}^{\sharp}$, fol. 46–47. $N = 18$; 14 are occultations of Aldebaran, 4 of Regulus.

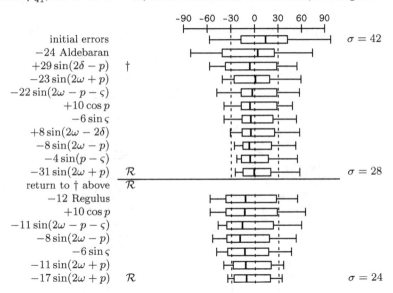

initial errors — $\sigma = 42$
-24 Aldebaran
$+29\sin(2\delta - p)$ $\dagger$
$-23\sin(2\omega + p)$
$-22\sin(2\omega - p - \varsigma)$
$+10\cos p$
$-6\sin\varsigma$
$+8\sin(2\omega - 2\delta)$
$-8\sin(2\omega - p)$
$-4\sin(p - \varsigma)$
$-31\sin(2\omega + p)$ $\mathcal{R}$ — $\sigma = 28$
return to $\dagger$ above $\mathcal{R}$
-12 Regulus
$+10\cos p$
$-11\sin(2\omega - p - \varsigma)$
$-8\sin(2\omega - p)$
$-6\sin\varsigma$
$-11\sin(2\omega + p)$
$-17\sin(2\omega + p)$ $\mathcal{R}$ — $\sigma = 24$

Cod. $\mu_{41}^{\sharp}$, fol. 342. $N = 14$, all solar eclipses. The narrowness of the box plots reflects the homogeneity of the data set. Note that during solar eclipses $\omega \approx 0$ and $\delta \approx 0$, which explains why argument ω is under-represented.

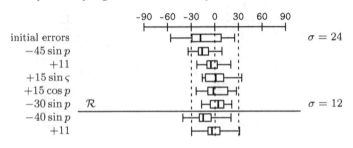

initial errors — $\sigma = 24$
$-45\sin p$
$+11$
$+15\sin\varsigma$
$+15\cos p$
$-30\sin p$ $\mathcal{R}$ — $\sigma = 12$
$-40\sin p$
$+11$

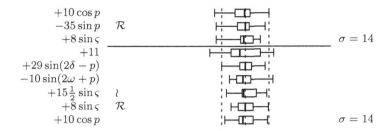

$$+10\cos p$$
$$-35\sin p \qquad \mathcal{R}$$
$$+8\sin\varsigma$$
$$+11$$
$$\sigma = 14$$
$$+29\sin(2\delta - p)$$
$$-10\sin(2\omega + p)$$
$$+15\tfrac{1}{2}\sin\varsigma \qquad \wr$$
$$+8\sin\varsigma \qquad \mathcal{R}$$
$$+10\cos p \qquad\qquad\qquad \sigma = 14$$

Cod. $\mu_{33}^{\sharp}$, sheet 1. $N = 55$, consisting of 14 solar eclipses and 41 occultations.

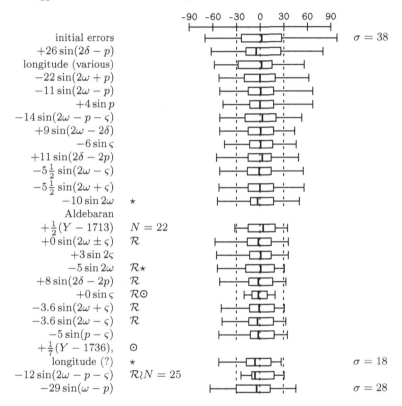

$$\text{initial errors} \qquad \sigma = 38$$
$$+26\sin(2\delta - p)$$
$$\text{longitude (various)}$$
$$-22\sin(2\omega + p)$$
$$-11\sin(2\omega - p)$$
$$+4\sin p$$
$$-14\sin(2\omega - p - \varsigma)$$
$$+9\sin(2\omega - 2\delta)$$
$$-6\sin\varsigma$$
$$+11\sin(2\delta - 2p)$$
$$-5\tfrac{1}{2}\sin(2\omega - \varsigma)$$
$$-5\tfrac{1}{2}\sin(2\omega + \varsigma)$$
$$-10\sin 2\omega \qquad \star$$
$$\text{Aldebaran}$$
$$+\tfrac{1}{2}(Y - 1713) \qquad N = 22$$
$$+0\sin(2\omega \pm \varsigma) \qquad \mathcal{R}$$
$$+3\sin 2\varsigma$$
$$-5\sin 2\omega \qquad \mathcal{R}\star$$
$$+8\sin(2\delta - 2p) \qquad \mathcal{R}$$
$$+0\sin\varsigma \qquad \mathcal{R}\odot$$
$$-3.6\sin(2\omega + \varsigma) \qquad \mathcal{R}$$
$$-3.6\sin(2\omega - \varsigma) \qquad \mathcal{R}$$
$$-5\sin(p - \varsigma)$$
$$+\tfrac{1}{7}(Y - 1736), \qquad \odot$$
$$\text{longitude (?)} \qquad \star \qquad \sigma = 18$$
$$-12\sin(2\omega - p - \varsigma) \qquad \mathcal{R}\wr N = 25 \qquad \sigma = 18$$
$$-29\sin(\omega - p) \qquad\qquad\qquad \sigma = 28$$

Cod. $\mu_{33}^{\sharp}$, sheet 2. $N = 90$: 29 eclipses and 61 occultations, see Fig. 8.4. The persistent large negative tails are caused by only two dates, which appear only on this sheet; without these outliers, the plots would be nearly symmetrical. Analysis of the final columns of this sheet was problematic.

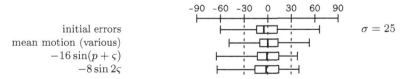

$$\text{initial errors} \qquad \sigma = 25$$
$$\text{mean motion (various)}$$
$$-16\sin(p + \varsigma)$$
$$-8\sin 2\varsigma$$

$+7\sin(2\omega - 2p)$		
$-10\sin 2p$		
$-10\sin(4\omega - 2p)$		
Apogee $+60$		
$-8\sin(\omega - p)$		
$+4\sin(2\omega + \varsigma)$	$\star$	
$+6\sin p$		
$-4\sin(2\omega + p)$		
$-7\sin 2p$	$\mathcal{R}$	
$-8\sin(4\omega - 2p)$	$\mathcal{R}$	
Regulus -5, other occ $+5.5\sin\varsigma$	$\star$	
$+3.5\sin(2\omega - 2p)$, Taurus occ extra -5	$\mathcal{R}$	
$-4\sin 2\varsigma$	$\mathcal{R}$	
$+2\sin(2\omega + \varsigma)$	$\mathcal{R}\star$	
$+5\sin(2\omega - 2p)$,	$\mathcal{R}$	
$-6\sin 2\varsigma$,	$\mathcal{R}$	
$+3\sin(2\omega + \varsigma)$	$\mathcal{R}$	
$+2$	$\odot\;\mathfrak{D}$	
unknown	(part $\mathfrak{l}$)	
$+8\sin(2 \cdot \mathrm{Long}_{\mathfrak{D}})$,	$\odot\;\mathfrak{D}\;\mathfrak{l}$	
longitude (various)	$\star$	
$+0\sin p$	$\mathcal{R}\mathfrak{l}$	$\sigma = 14$
$+8\sin(2\omega - 2\varsigma)$	$\star$	
$-10\sin(4\omega - 2p)$	$\mathcal{R}$	
$+3\sin\varsigma$	$\mathcal{R}\star$	
$-4\sin 4\omega$	$\star$	
$-14\sin(p + \varsigma)$	$\mathfrak{l}$	$\sigma = 15$
$-16\sin(2\omega + 2p)$	$\star\;\mathfrak{l}$	
not understood	$\mathfrak{l}N = 22$	
$-8\sin 4\omega$	$\mathcal{R}\star$	
$-15\sin(2\omega - p - \varsigma)$	$\star$	
not understood	$\star$	

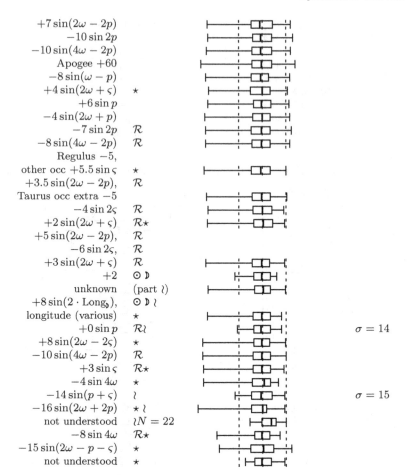

Cod. $\mu_{33}^{\sharp}$, sheet 3. $N = 56$; there are 11 eclipses and 45 occultations. It is interesting to see the tails move to the left, while the central boxes remain almost untouched.

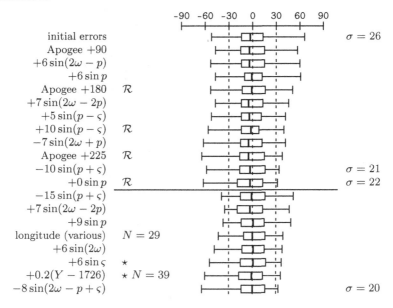

Cod. $\mu_{33}^{\sharp}$, sheet 4. $N = 91$.

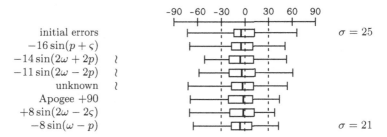

Cod. $\mu_{33}^{\sharp}$, sheet 5. $N = 80$; 25 eclipses and 55 occultations.

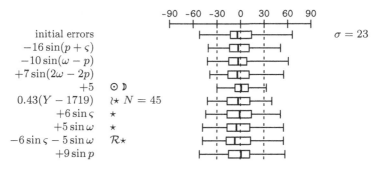

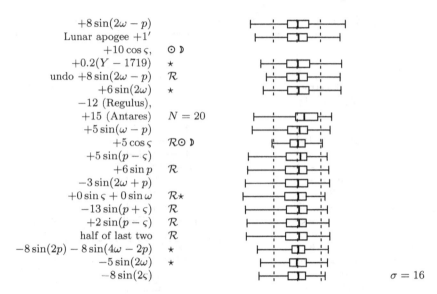

$+8\sin(2\omega - p)$

Lunar apogee $+1'$

$+10\cos\varsigma,$ ⊙ ☽

$+0.2(Y - 1719)$ ⋆

undo $+8\sin(2\omega - p)$ $\mathcal{R}$

$+6\sin(2\omega)$ ⋆

-12 (Regulus),

$+15$ (Antares) $N = 20$

$+5\sin(\omega - p)$

$+5\cos\varsigma$ $\mathcal{R}$⊙ ☽

$+5\sin(p - \varsigma)$

$+6\sin p$ $\mathcal{R}$

$-3\sin(2\omega + p)$

$+0\sin\varsigma + 0\sin\omega$ $\mathcal{R}$⋆

$-13\sin(p + \varsigma)$ $\mathcal{R}$

$+2\sin(p - \varsigma)$ $\mathcal{R}$

half of last two $\mathcal{R}$

$-8\sin(2p) - 8\sin(4\omega - 2p)$ ⋆

$-5\sin(2\omega)$ ⋆

$-8\sin(2\varsigma)$ $\sigma = 16$

Cod. $\mu_{33}^{\sharp}$, sheet 6. The 27 lunar eclipses (*top*) and 14 Aldebaran occultations (*bottom*) are presented separately.

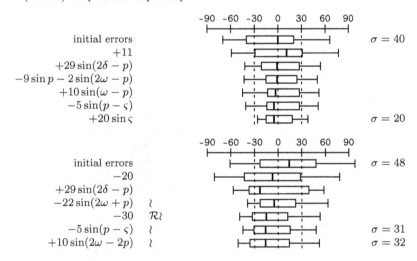

initial errors $\sigma = 40$

$+11$

$+29\sin(2\delta - p)$

$-9\sin p - 2\sin(2\omega - p)$

$+10\sin(\omega - p)$

$-5\sin(p - \varsigma)$

$+20\sin\varsigma$ $\sigma = 20$

initial errors $\sigma = 48$

-20

$+29\sin(2\delta - p)$

$-22\sin(2\omega + p)$ ☽

-30 $\mathcal{R}$☽

$-5\sin(p - \varsigma)$ ☽

$+10\sin(2\omega - 2p)$ ☽ $\sigma = 31$

$\sigma = 32$

Cod. $\mu_{33}^{\sharp}$, sheet 7. $N = 18$, see Fig. 8.3. Initially 14 occultations of Aldebaran. With the new start, 4 Regulus occultations were added. Initial errors are not explicitly listed on the sheet, but are, after the new start, continued from Cod. $\mu_{41}^{\sharp}$, fol. 46–47.

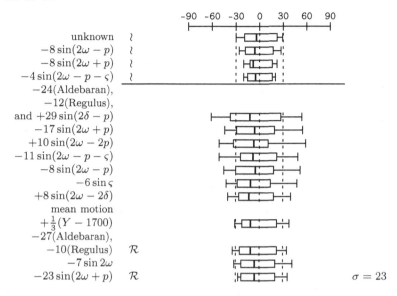

Cod. $\mu_{33}^{\sharp}$, sheet 8. $N = 14$, solar eclipses. Just as on the spreadsheet of Cod. $\mu_{41}^{\sharp}$ fol. 342, the homogeneous data yield narrow box plots.

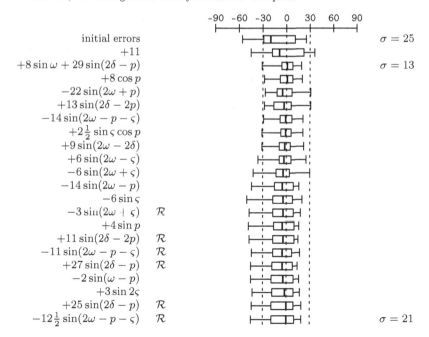

Display B.1 All amendments sorted by equation, columns in approximate chronological order (continued on next page)

eqn.	arg.	sheet P44	sheet P47		sheet 7		sheet P342			sheet 6	
I	$\varsigma, 2\varsigma$		$-\frac{1}{120}$	$-\frac{1}{120}$	$-\frac{1}{120}$		$+15''$	$+8''$	$+15\frac{1}{2}'',$ $+8''$	$+20''$	
II	$2\omega + \varsigma$										
III	$2\omega - \varsigma$										
IV	$2\omega - p + \varsigma$	$+\frac{1}{6},$ $+\frac{1}{12}$									
V	$2\omega - p - \varsigma$		$-\frac{3}{10}$	$-\frac{3}{20}$	$-\frac{3}{10}$	$-\frac{3}{20}$					
VI	$2\omega + p$	$-\frac{1}{6}$	$-\frac{1}{4},$ $-\frac{1}{3}$	$-\frac{1}{8},$ $-\frac{3}{16}$	$-\frac{1}{12}$	$-\frac{3}{16},$ $-\frac{1}{4}$		$-10''$			$-\frac{1}{4}$
VII	$2\delta - p$	$+\frac{1}{2}$	$+\frac{1}{2}$	$+\frac{1}{2}$	$+\frac{1}{2}$				$+\frac{1}{2}$	$+\frac{1}{2}$	$+\frac{1}{2}$
VIII	$p - \varsigma$	$-\frac{1}{2}$	$-\frac{1}{10}$							$-\frac{1}{8}$	$-\frac{1}{8}$
IX	$2\omega - 2\delta$	$+\frac{1}{5}$	$+\frac{1}{6}$			$+\frac{1}{6}$					
X	$\omega - p,$ $2\omega - 2p$					$+\frac{1}{20}$				$+\frac{1}{20}$	$+\frac{1}{20}$
XI	$p, 2p, 3p$						$-45'',$ $-30''$	$-40'',$ $-35''$		$+\frac{1.5}{60\cdot60}$	
XII	$2\omega - p,$ $4\omega - 2p$		$+\frac{6}{60\cdot60}$	$+\frac{6}{60\cdot60}$	$+\frac{6}{60\cdot60}$	$+\frac{6}{60\cdot60}$				$+\frac{1.5}{60\cdot60}$	
XIII	$\omega, 2\omega, 3\omega,$ 4ω	$+20''$			$-7''$						
XIV	2δ										
	long.	$-19'',$ $-34'',$ $-30''$	$-24''$	$-24''$	various		$+11''$	$+11''$	$+11''$	$+11''$	$-20'',$ $-30''$
	mot.med.				$+10''$ per 30 year						
	Apog (sign?)		$+90''$	$+90''$			$+135''$	$+90''$	$+90''$		
	$p + \varsigma$										
	$2\delta - 2p$										
	$2\omega - 2\varsigma$										
	$2\omega + 2p$										
	etc		$2\omega - 2p$ $+ \varsigma$ nihil								

Display B.1 All amendments sorted by equation, columns in approximate chronological order (continued from previous page)

eqn.	arg.	sheet 8	sheet 1	sheet 3	sheet 5	sheet 2	sheet 4
I	ς, 2ς	$-\frac{1}{120}$, $-\frac{1}{120}$ $+3''[2ς]$	$-\frac{1}{120}$, $-\frac{1}{120}$ $+3''[2ς]$, 0	$+\frac{1}{120}$	$+\frac{1}{120}$, $+\frac{1}{120}-8''[2ς]$	$-8''[2ς]$, $-8''[2ς]+\frac{1}{120}$, $-4''[2ς]+\frac{1}{120}$, $-6''[2ς]+\frac{1}{120}$, $-6''[2ς]+\frac{1}{240}$	
II	$2\omega + ς$	$+\frac{1}{10}$, $+\frac{1}{20}$	$+\frac{1}{10}$, 0, $+\frac{1}{15}$			$-\frac{1}{14}$, $-\frac{1}{28}$, $-\frac{3}{56}$	
III	$2\omega - ς$	$-\frac{1}{11}$	$+\frac{1}{11}$, 0, $+\frac{1}{17}$				
IV	$2\omega - p + ς$			$-\frac{1}{15}$			
V	$2\omega - p - ς$	$-\frac{1}{5}$, $-\frac{1}{5}+\frac{1}{25}$, $-\frac{1}{5}+\frac{1}{25}-\frac{1}{50}$	$-\frac{1}{5}$, $-\frac{1}{6}$			$-\frac{1}{5}$	
VI	$2\omega + p$	$-\frac{1}{4}$	$-\frac{1}{4}$	$-\frac{1}{9}$	$-\frac{1}{20}$	$-\frac{1}{17}$	
VII	$2\delta - p$	$+\frac{1}{2}$, $+\frac{1}{2}-\frac{1}{30}$, $+\frac{1}{2}-\frac{1}{15}$	$+\frac{1}{2}-\frac{1}{20}$				
VIII	$p - ς$		$-\frac{1}{8}$	$+\frac{1}{6}$, $+\frac{1}{3}$	$+\frac{1}{6}$, $+\frac{1}{12}$, $+\frac{1}{8}$		
IX	$2\omega - 2\delta$	$+\frac{1}{5}$	$+\frac{1}{5}$				
X	$\omega - p$, $2\omega - 2p$	$-2''[\omega - p]$	$+29''[\omega - p]$	$+\frac{1}{30}$	$+\frac{1}{30}$, $+\frac{1}{30}+5''[\omega - p]$, $+\frac{1}{60}$, $+\frac{1}{45}$	$+\frac{1}{30}$, $+\frac{1}{30}-8''[\omega - p]$	$-\frac{1}{20}$, $-8''[\omega - p]$
XI	p, $2p$, $3p$	$-4''$	$-4''$	$-\frac{1}{60\cdot 60}$, 0	$-\frac{1.5}{60\cdot 60}$, $-9''$, $-6''$, $-8''[2p]$	$-10''[2p]$, $-10''[2p]-\frac{1}{60\cdot 60}$, $-10''[2p]$, $-7''[2p]$	
XII	$2\omega - p$, $4\omega - 2p$	$+\frac{10}{60\cdot 60}$	$+\frac{8}{60\cdot 60}$	$-\frac{6}{60\cdot 60}$	$-\frac{6}{60\cdot 60}$, 0, $-8''[4\omega - 2p]$	$-10''[4\omega - 2p]$, $-8''[4\omega - 2p]$, $-10''[4\omega - 2p]$	
XIII	ω, 2ω, 3ω, 4ω	$+8''$	$-\frac{1}{4\cdot 60}$, $-\frac{1}{8\cdot 60}$	$-\frac{1}{7\cdot 60}$	$+5\frac{1}{2}''[\omega]$, $-5\frac{1}{2}''[\omega]$, $0[\omega]$, $-\frac{1}{7\cdot 60}$	$-4''[4\omega]$, $-8''[4\omega]$	
XIV	2δ						
	long.	$+11''$	$+11''$ etc.	var	various	various	
	mot.med.		ca. $+\frac{1}{2}$, $+\frac{9}{00}$	$+1''$ per 5 year	ca. $+0.4$, $+0.6$	$+4''$ per 25 year	
	Apog (sign?)	$+72''$		$+90''$, $+3'$, $+3'45''$	$+60''$, $+90$, $+45$	$+60''$	$+90''$
	$p + ς$			$-10''$	$-15''$, $-16''$, $-13''$, $-15\frac{1}{2}$	$-16''$, $-14''$, $+5''$	$-16''$
	$2\delta - 2p$	$+13''$, $+11$	$+11''$, $+8''$				
	$2\omega - 2ς$					$+8''$	$+8''$
	$2\omega + 2p$					$-16''$	$-14''$
	etc	$A - \frac{1}{30}$				$+8''[2Long]$	

Appendix C
Manuscript Sources

Manuscripts in the Niedersächsische Staats- und Universitäts-Bibliothek, Göttingen After Mayer's death, his widow sold the bulk of his papers to the Hanoverian government. They remained neglected partly in the observatory and partly in the archives of the Göttingen scientific society. Lichtenberg accessed the papers from 1773 onwards. He catalogued them and planned an edition of the most interesting parts of Mayer's unpublished writings. However, Lichtenberg realized only one volume of what was intended as a series.[1] After 24 years he returned the manuscript papers to the Academy. When a new astronomical observatory was erected outside the old city walls, they were transferred to that place. In 1965, they were relocated to the Niedersächsische Staats- und Universitäts-Bibliothek in Göttingen where they remain now.[2]

For these manuscripts I first give the abbreviated reference used in this book, then the official reference according to Meyer (1894), together with a descriptive title. These titles in German were either supplied by Mayer or they were descriptions drawn up by Lichtenberg; they usually provide a fair indication of the contents of the papers.[3]

Cod. μ_1	(Cod. Ms. T. Mayer 1) Eigene und fremde Beobachtungen von Stellen des Mondes und darüber angestellte Berechnungen.
Cod. μ_2	(Cod. Ms. T. Mayer 2) Analytische Rechnungen nebst Anwendung auf die Astronomie.
Cod. μ_3	(Cod. Ms. T. Mayer 3) Analytische Rechnungen nebst Anwendungen auf astronomische Untersuchungen.
Cod. μ_4	(Cod. Ms. T. Mayer 4) Rechnungen die besonders zur Mondtheorie gehören und Beobachtungen und Rechnungen von Stellen von Fixsternen.
Cod. μ_5	(Cod. Ms. T. Mayer 5) Berechnungen über den Mond zur Prüfung der Mondtafeln Mayer's.

---

[1] Mayer (1775).

[2] See also Cod. μ_1 (p. 1), Forbes (1971c, p. 17), Meyer (1894, p. 154).

[3] The index of Mayer's papers in Meyer (1894) has been reproduced in Forbes (1980, pp. 228–232); also see Zinner (1925).

Cod. μ_6 (Cod. Ms. T. Mayer 6) T. Mayeri Observationes astronomicae a° 1751–1756; von Bl. 35 an im Jahre 1756 (vgl. Mayer 15_{25}).

Cod. μ_7 (Cod. Ms. T. Mayer 7) Berechnungen und Beobachtungen, meist von Fixsternen.

Cod. μ_8 (Cod. Ms. T. Mayer 8) Berechnungen zur Mayer'schen Theorie des Mondes und des Magnets.

Cod. μ_{10} (Cod. Ms. T. Mayer 10) Sammlung verschiedener fremden und eignen Beobachtungen von Länge und Breite und von Beobachtungen zur Bestimmung der Mondflecken (Bl. 32–50 'Formeln für die Trigonometrie'; Bl. 54–89 'Bestimmung der Neigung des Mondaequators gegen die Bahn desselben', 63 §§).

Cod. μ_{12} (Cod. Ms. T. Mayer 12) 'Collectanea Mathematica 1747–53': Excerpte, fast nur Astron. Inhalts; darin S. 73 Brief von Chirstoph Maire S.J. an Lowiz, Rome 14 Sept. 1748, Copie.

Cod. μ_{14} (Cod. Ms. T. Mayer 14) Beobachtungen von Finsternissen 1749–53 (Bl. 21–23 von Lowiz 1753); theilweise gedruckt in den Comm. Soc. Gott. 1751.

Cod. $\mu_2^\sharp$ (Cod. Ms. T. Mayer 15_2) Briefe an T. Mayer [details omitted].

Cod. $\mu_6^\sharp$ (Cod. Ms. T. Mayer 15_6) Lehre von der Parallaxe des Mondes.

Cod. $\mu_{11}^\sharp$ (Cod. Ms. T. Mayer 15_{11}) Collectanea geographica et mathematica 1747. (Etwa 300 Bl.). Grösstentheils Excerpte; am Schluss Untersuchungen über die geographische Länge und Breite der Stadt Nürnberg und von den Construction der Land-Karten (unvollständig).

Cod. $\mu_{14}^\sharp$ (Cod. Ms. T. Mayer 15_{14}) Tabulae solares et lunae (30 Bl.), nicht die in den Commentationes 1752 gedruckten. Vgl. 15 n. 15 und 49.

Cod. $\mu_{15}^\sharp$ (Cod. Ms. T. Mayer 15_{15}) Mondtafeln (wahrscheinlich älterer Entwurf).

Cod. $\mu_{28}^\sharp$ (Cod. Ms. T. Mayer 15_{28}) Rechnungen zur Mondtheorie, 3 Hefte, das letzte 1755 Jan. und Febr. (etwa 120 Bl.); vgl. 15 n. 30.

Cod. $\mu_{30}^\sharp$ (Cod. Ms. T. Mayer 15_{30}) Theoria Lunae, etwa 80 Bl. mit Rechnungen, vgl. 15 n. 28.

Cod. $\mu_{33}^\sharp$ (Cod. Ms. T. Mayer 15_{33}) Berechnung der Mondbeobachtungen, Vergleichung mit der Theorie (etwa 30 Bl.).

Cod. $\mu_{41}^\sharp$ (Cod. Ms. T. Mayer 15_{41}) Etwa 300 Bl. über Sternbedeckungen und Mondbeobachtungen, frühere und eigene Beobachtungen und Rechnungen.

Cod. $\mu_{48}^\sharp$ (Cod. Ms. T. Mayer 15_{48}) Historia Eclipsium ab anno 1610, quo ad telescopio observationes fieri coeperunt (bis 1678, Excerpte).

Cod. $\mu_{49}^\sharp$ (Cod. Ms. T. Mayer 15_{49}) Sonnen- und Mondtafeln, vgl. 15 n. 14.

Cod. $\mu_{54}^\sharp$ (Cod. Ms. T. Mayer 15_{54}) Astronomische Rechnungen, etwa 20 Bl.

Cod. $\mu_{55}^\sharp$ (Cod. Ms. T. Mayer 15_{55}) Etwa 400 Bl., ungeordnet, zum Theil Bruchstücke mathematischen und astronomischen Inhalts von Collegheften, Aufsätzen und dergl., zum Theil Beobachtungen und Rechnungen.

Michael. 320   (Cod. Ms. Michael. 320) Briefe an und von Michaelis.

Philos. 159   (Cod. Ms. Philos 159) Briefe von und an J. Tobias Mayer.

Manuscripts in the Cambridge University Library, Cambridge The archives of the Royal Greenwich Observatory were transferred to Cambridge University Library in 1990. Further information on these archives is available online via the Janus electronic catalogue of the library, http://www.lib.cam.ac.uk/.

Royal Greenwich Observatory Archives, Papers of Edmond Halley, RGO 2

RGO 2/14   Printed lunar tables.

Royal Greenwich Observatory Archives, Papers of James Bradley and Nathaniel Bliss, RGO 3

RGO 3/33   Lunar calculations.
RGO 3/41   Computations for the Moon.
RGO 3/42   Tables.
RGO 3/43   Observations, calculations and correspondence.

Royal Greenwich Observatory Archives, Papers of Nevil Maskelyne, RGO 4

RGO 4/67   Stellar observations and calculations.
RGO 4/80   Moon's observed place.
RGO 4/81   Moon's observed place.
RGO 4/84   Mason's lunar tables.
RGO 4/108   Lunar theory and the Newtonian system.
RGO 4/125   Solar and lunar tables.
RGO 4/130   Index to some solar tables.
RGO 4/140   Calculations in trigonometry and central forces.
RGO 4/149   Letters from Maskelyne to Henry Andrews.
RGO 4/187   Miscellaneous correspondence.
RGO 4/193   Lunar calculations.
RGO 4/202   Ephemerides.
RGO 4/211   Papers on astronomy.
RGO 4/282   Principles of navigation.
RGO 4/320   Miscellaneous papers.

Papers in Edinburgh University Library, Edinburgh The papers of the late Professor Eric Gray Forbes have been transferred to the Special Collections Division of Edinburgh University Library.

GB 237 Coll-132, folder 5: Tobias Mayer

8   The astronomical correspondence between Joseph Nicholas de l'Isle and Tobias Mayer
12   Tobias Mayer's Mathematical Atlas

References

Airy GB (1834) Gravitation: an elementary explanation of the principal perturbations in the solar system. Charles Knight, London; 2nd edn: Macmillan, London (1884)

Andrewes WJH (ed) (1996) The quest for longitude: the proceedings of the longitude symposium. Collection of historical scientific instruments. Harvard University Press, Cambridge, MA

Auwers A (1894) Tobias Mayer's Sternverzeichniss: nach den Beobachtungen auf der Göttinger Sternwarte in den Jahren 1756 bis 1760. Neu bearbeitet von Arthur Auwers. Engelmann, Leipzig

Baasner R (1987) Das Lob der Sternkunst: Astronomie in der deutschen Aufklärung. Vandenhoeck & Ruprecht, Göttingen

Baily F (ed) (1835) An account of the Revd. John Flamsteed, the first Astronomer-Royal: compiled from his own manuscripts, and other authentic documents, never before published, London; Reprinted Dawson, London (1966)

Beer A, Beer P (eds) (1976) The origins, achievement and influence of the Royal Observatory, Greenwich: 1675–1975. In: Proceedings of the symposium held at the national maritime museum, Greenwich, 13–18 July 1975, vistas in astronomy, vol. 20

Bohnenberger J (1795) Anleitung zur Geographischen Ortsbestimmung vorzüglich vermittelst des Spiegelsextanten. Vandenhoeck und Ruprecht, Göttingen; Revised by GA Jahn (1852)

Bradley J (1748) A letter to the right honourable george earl of macclesfield concerning an apparent motion observed in some of the fixed stars. Philos Trans (1683–1775) 45:1–43

Brouwer DJ (1864) Theoretische en Praktische Zeevaartkunde, benevens eene beknopte Verhandeling over de Hydrographie, Nieuwediep, 2 vols

Brouwer D, Clemence GM (1961) Methods of celestial mechanics. Academic, New York

Brown EW (1896) An introductory treatise on the lunar theory. Cambridge University Press, Cambridge

Buchwald JZ (2006) Discrepant measurements and experimental knowledge in the early modern era. Arch Hist Exact Sci 60:565–649

Buchwald JZ, Cohen IB (eds) (2001) Isaac Newton's natural philosophy. Dibner Institute studies in the history of science and technology, MIT, Cambridge, MA

Campbell-Kelly M et al. (eds) (2003) The history of mathematical tables: from sumer to spreadsheets. Oxford University Press, Oxford

Chandler B (1996) Longitude in the context of mathematics. In: The quest for longitude: the proceedings of the longitude symposium, pp 34–42

Chapin SL (1978) Lalande and the longitude: a little known London voyage of 1763. Notes Rec R Soc London 32:165–180

Chapman A (1982) Three North country astronomers. Richardson, Manchester

Chapront-Touzé M, Chapront J (1983) The lunar ephemeris ELP2000. Astron Astrophys 124:50–62

Chauvenet W (1863) A manual of spherical and practical astronomy, 2 vols. Lippincott, Philadelphia, PA; 5th edn: (1891)

Clairaut AC (1752a) De l'orbite de la lune, en ne négligeant pas les quarrés des quantités de même ordre que les forces perturbatrices. Histoire de l'Académie Royale des Sciences avec les mémoires de mathématique et de physique tirés des registres de cette Académie (année 1748), pp 421–440

Clairaut AC (1752b) Théorie de la lune: déduite du seul principe de l'attraction réciproquement proportionnelle aux quarrés des distances. l'Académie Impériale des Sciences, St. Petersbourg; 2nd enlarged edn: Dessaint, Paris (1765)

Clairaut AC (1758) Lettre de Monsieur Clairaut à Messieurs les Auteurs du Journal des Sçavans. Journal des Sçavans 67–82

Cook A (1988) The motion of the moon. Adam Hilger, Bristol

Cook A (1998) Edmond Halley: charting the heavens and the seas. Clarendon Press, Oxford

Cook A (2000) Success and failure in Newton's lunar theory. Astron Geophys 41(6):21–25

Cotter CH (1968) A history of nautical astronomy. Hollis & Carter, London

d'Alembert J (1754–1756) Recherches sur différens points importans du système du monde. Chez David l'aîné, Paris, 3 vols; Facsimile reprint: Culture et civilisation, Bruxelles (1966)

d'Alembert J (1761–1780) Opuscules mathématiques, ou mémoires sur différens sujets de géométrie, de méchanique, d'optique, d'astronomie &c. Chez David l'aîné, Paris, 8 vols.

Dalen Bv (1993) Ancient and medieval astronomical tables: mathematical structure and parameter values. PhD thesis, Utrecht University

Davids CA (1985) Zeewezen en Wetenschap: De wetenschap en de ontwikkeling van de navigatietechniek in Nederland tussen 1585 en 1815. De Bataafsche Leeuw, Amsterdam/Dieren

Delambre JBJ (1806) Tables Astronomiques publiées par le Bureau des Longitudes de France. Première partie. Tables du Soleil par M. Delambre. Tables de la Lune par M. Bürg. Paris

Delambre JBJ (1827) Histoire de l'Astronomie au dix-huitième Siècle. Bachelier, Parijs

Euler L (1746) Tabulae astronomicae Solis et Lunae. In: Opuscula varii argumenti, vol 1. Berlin, pp 137–168; Reprinted in Opera Omnia II, 23, pp 1–10

Euler L (1749a) Recherches sur la question des inégalités du mouvement de Saturne et de Jupiter (Pièce qui a remporté le prix de l'Académie royale des sciences en 1748 Sur les inégalités du mouvement de Saturne & de Jupiter). Martin, Coignard, and Guerin, Paris; Reprinted in Opera Omnia II, 25, pp 45–157

Euler L (1749b) Recherches sur le mouvement des corps célestes en général. Mémoires de l'académie des sciences de Berlin 3:93–143; Reprinted in Opera Omnia II, 25, pp 1–44

Euler L (1753) Theoria motus lunae: exhibens omnes eius inaequalitates.... Petropolitanae: Impensis Academiae Imperialis Scientiarum; Reprinted in Opera Omnia II, 23, pp 64–336

Euler L (1770a) Nouvelle méthode de déterminer les dérangemens dans le mouvement des corps célestes, causés par leur action mutuelle. Histoire de l'Académie Royale des Sciences et des Belles-Lettres de Berlin 19:141–179; Reprinted in Opera Omnia II, 24

Euler L (1770b) Réflexions sur les diverses manieres dont on peut représenter le mouvement de la lune. Histoire de l'Académie Royale des Sciences et des Belles-Lettres de Berlin 19:180–193; Reprinted in Opera Omnia II, 26

Euler L (1772) Theoria motuum lunae nova methodo pertractata una cum tabulis astronomicis.... Academiae Imperialis Scientiarum, Petropoli; Reprinted in Opera Omnia II, 22

Farebrother RW (1998) Fitting linear relationships, a history of the calculus of observations 1750 – 1900. Springer, New York

Flamsteed J (1725) Historia coelestis britannica. H Meere, London

Forbes EG (1965) The foundations and early development of the nautical almanac. J Inst Navigation 18:391–401

Forbes EG (1966) Tobias Mayer's lunar tables. Ann Sci 22:105–116

Forbes EG (1971a) The Euler–Mayer correspondence (1751–55): a new perspective on eighteenth-century advances in the lunar theory. Macmillan, London

Forbes EG (1971b) Tobias Mayer's new astrolabe (1759): its principles and construction. Ann Sci 27:109–116

Forbes EG (1971c) Tobias Mayer's opera inedita, the first translation of the Lichtenberg edition of 1775. American Elsevier, New York

Forbes EG (1972) The unpublished Writings of Tobias Mayer. Arbeiten aus der niedersächsischen Staats- und Universitätsbibliothek Göttingen; Bd. 9–11. Vandenhoeck & Ruprecht, Göttingen, 3 vols.; 1: Astronomy and Geometry, 2: Artillery and mechanics, 3: The Theory of the Magnet and its application to terrestrial magnetism

Forbes EG (1974) Tobias Mayer's debt to Leonhard Euler. In: Actes du XIIIe Congrès International d'Histoire des Sciences, vol 6, pp 295–299

Forbes EG (1975) Greenwich Observatory, The Royal Observatory at Greenwich and Herstmonceux, 1675–1975, vol 1. Origins and early history (1675–1835). Taylor and Francis, London

Forbes EG (1980) Tobias Mayer (1723–62): pioneer of enlightened science in Germany. Vandenhoeck & Ruprecht, Göttingen

Forbes EG (1983) La correspondance astronomique entre Joseph–Nicholas Delisle et Tobias Mayer. Revue d'Histoire des Sciences 36:113–151

Forbes EG, Gapaillard J (1996) The astronomical correspondence between abbé de Lacaille and Tobias Mayer. Revue d'Histoire des Sciences 49(4):483–542

Forbes EG, Wilson CA (1995) The solar tables of Lacaille and the lunar tables of Mayer. In: Planetary astronomy from the renaissance to the rise of astrophysics: the eighteenth and nineteenth centuries, Chap 18

Frisius P (1768) De gravitate universali corporum libritres. Milan

Gaab H (2001) Johann Gabriel Doppelmayr (1677–1750). Beiträge zur Astronomiegeschichte Band 4:46–99

Gautier A (1817) Essai historique sur le Problème des Trois Corps. Veuve Courcier, Paris

Gaythorpe SB (1957) Jeremiah Horrox and his 'New Theory of the Moon'. J Br Astron Assoc 67(4):134–144

Giorgini JD, Yeomans DK, Chamberlin AB, Chodas PW, Jacobson RA, Keesey MS, Lieske JH, Ostro SJ, Standish EM, Wimberly RN (1996) JPL's on-line solar system data service. Bull Am Astron Soc 28(3):1158

Godfray H (1852) An elementary treatise on the lunar theory. Macmillan, Cambridge

Golland LA, Golland RW (1993) Euler's troublesome series: an early example of the use of trigonometric series. Historia Mathematica 20:54–67

Gowing R (1983) Roger Cotes: natural philosopher. Cambridge University Press, Cambridge

Graff K (1914) Grundriß der geographischen Ortsbestimmung aus astronomischen Beobachtungen. G.J. Göschen, Berlin

Gutzwiller MC (1998) Moon–Earth–Sun: the oldest three-body problem. Rev Mod Phys 70(2):589–639

Guyou E (1902) La méthode des distances lunaires. La Revue Maritime et Coloniale, pp 943–963

Hald A (1990) A history of probability and statistics and their applications before 1750. Wiley, New York

Hansen T (1964) Arabia Felix : the Danish expedition of 1761-1767. Collins, London

Hell M, Pilgram A (1772) Ephemerides astronomicae ad meridianum Vindobonensem calculis definitae. Vienna

Hellman CD (1932) John Bird (1709-1776): mathematical instrument-maker in the strand. Isis 17:127–153

Horrocks J (1673) Jeremiae Horroccii Opera posthuma. London

Howse D (1980) Greenwich time and the discovery of the longitude. Oxford University Press, Oxford

Howse D (1989) Nevil Maskelyne: The seaman's astronomer. Cambridge University Press, Cambridge

Jordan DW (1885) Grundzüge der Astronomischen Zeit- und Ortsbestimmung. Julius Springer, Berlin

Jørgensen NT (1974) On the moon's elliptic inequality, evection, and variation and Horrox's 'New Theory of the Moon'. Centaurus 18:316–319

Juškevič A (ed) (1975) Leonhardi Euleri commercium epistolicum. Opera Omnia IV, 1, Birkhauser, Basle

Katz VJ (1987) The calculus of the trigonometric functions. Historia Mathematica 14: 311–324

Kollerstrom N (2000) Newton's forgotten lunar theory: his contribution to the quest for longitude. Green Lion Press, Santa Fe

Lacaille NLd (1759) Méthode pour trouver facilement les Longitudes en mer par le moyen de la Lune. In: Connoissance des Temps pour l'Année 1761, Paris, pp 174–193

Lagrange JL (1882) Oeuvres de Lagrange: publ. par les soins de J.-A. Serret; sous les auspices du Ministre de l'Instruction Publique, vol 13. Gauthier-Villars, Paris

Lalande JJ (1764) Astronomie. Veuve Dessaint, 2nd edn, Paris, 1771

Laplace PS (1802) Traité de mécanique céleste, vol 3. Duprat, Paris

Lémery M (1780) Comparaison de cinq cents vingt-cinq Observations de la Lune, avec des Tables de Clairaut, celle de Mayer & celles qui ont été corrigées en Angleterre pour le Nautical Almanach. In: Connoissance des Temps pour l'Année 1783, Paris, pp 352–371

Lemonnier PC, Keill J (1746) Institutions astronomiques ou Leçons élémentaires d'astronomie [traduites du latin de Jean Keill et augmentées par Pierre-Charles Lemonnier]. Guerin, Paris

Linton CM (2004) From Eudoxus to Einstein: a history of mathematical astronomy. Cambridge University Press, Cambridge

Lowitz GM (1747) Le Monde eclipsé ou Représentation géographique de l'éclipse de la terre ou du soleil qui arrivera le 25 Iuillet 1748. Homann, Nuremberg

Malin SR, Roy AE, Beer P (eds) (1985) Longitude zero, 1884–1984: Proceedings of an international symposium held at the National Maritime Museum, Greenwich, London, 9–13 July 1984 to mark the centenary of the adoption of the Greenwich Meridian, vistas in astronomy, vol. 28

Marguet F (1931) Histoire Générale de la Navigation du XVe au XXe siècle. Société d'Editions Géographiques, Maritimes et Coloniales, Paris

Maskelyne N (1761) A letter from the Rev Nevil Maskelyne ... to the Rev Thomas Birch ... containing the results of observations of the distance of the moon from the sun and fixed stars, made in a voyage from England to the Island of St Helena in order to determine the longitude of the ship, from time to time; together with the whole process of computation used on this occasion. Philos Trans R Soc 52:558–577

Maskelyne N (1763) The British Mariner's Guide containing ... Instructions for the Discovery of the Longitude ... by observations of the distance of the moon from the sun and stars, taken with Hadley's Quadrant. London

Mason C (1787) Mayer's tables improved by Mr. Charles Mason. Commisioners of Longitude, London

May WE (1952) Navigational accuracy in the eighteenth century. J Inst Navigation 5:71–73

May WE (1973) A history of marine navigation. G T Foulis, Henley-on-Thames

Mayer T (1745) Mathematischer Atlas, in welchem auf 60 Tabellen alle Theile der Mathematic vorgestellet und nicht allein überhaupt zu bequemer Wiederholung, sondern auch den Anfängern besonders zur Aufmunterung durch deutliche Beschreibung u. Figuren entworfen werden. Pfeffel, Augspurg

Mayer T (1748) Représentation de l'eclipse partiale de la Lune qui arrivera la nuit du 8 au 9 d'août, 1748 dressée suivant les lois de la vraie projection et accompagné d'un imprimé qui lui servira d'éclaircissement. Homann, Nuremberg

Mayer T (1750a) Abhandlung über die Umwälzung des Monds um seine Axe und die scheinbare Bewegung der Mondsflecken; worinnen der Grund einer verbesserten Mondsbeschreibung aus neuen Beobachtungen geleget wird. Erster Theil. In: Kosmographische Nachrichten und Sammlungen auf das Jahr 1748: zum Wachsthume der Weltbeschreibungswissenschaft von den Mitgliedern der Kosmographischen Gesellschaft zusammengetragen, pp 52–183

Mayer T (1750b) Astronomische Beobachtung der grossen Sonnenfinsterniss J.J. 1748 den 25 Jul. zu Nürnberg in dem homännischen Hause angestellet. Mit nöthigen Anmerkungen. In: Kosmographische Nachrichten und Sammlungen auf das Jahr 1748: zum Wachsthume der Weltbeschreibungswissenschaft von den Mitgliedern der Kosmographischen Gesellschaft zusammengetragen, pp 11–40

Mayer T (1750c) Germaniae atque in ea locorum principaliorum Mappa Critica, ex latitudinum observationibus, quas hactenus colligere licuit, omnibus; mappis specialibus compluribus; itinerariis antiquis Antonini, Augustano et Hierosolymitano, adhibita circumspectione ac saniori crisi concinnata simulque cum aliorum Geographorum mappis comparata a Tob. Mayero.

Mayer T (1750d) Tobias Mayers Beweis daß der Mond keinen Luftkreis habe. In: Kosmographische Nachrichten und Sammlungen auf das Jahr 1748: zum Wachsthume der Weltbeschreibungswissenschaft von den Mitgliedern der Kosmographischen Gesellschaft zusammengetragen, pp 397–419

Mayer T (1752) Latitudo Geographica Urbis Norimbergae e Novis Observationibus Deducta. Commentarii Societatis Regiae Scientiarum Gottingensis 1:373–378

Mayer T (1753a) In Parallaxin Lunae eiusdemque a Terra Distantiam Inquistio. Commentarii Societatis Regiae Scientiarum Gottingensis 2:159–182

Mayer T (1753b) Novae Tabulae motuum Solis et Lunae. Commentarii Societatis Regiae Scientiarum Gottingensis 2:383–431

Mayer T (1754) Tabularum Lunarium in Commentt. S. R. Tom. II. Contentarum Usus in Investiganda Longitudine Maris. Commentarii Societatis Regiae Scientiarum Gottingensis 3:356–397

Mayer T (1755) Experimenta circa Visus Aciem. Commentarii Societatis Regiae Scientiarum Gottingensis 4:120–135

Mayer T (1767) Theoria Lunæ juxta Systema Newtonianum. Commisioners of Longitude, London

Mayer T (1770) Tabulæ Motuum Solis et Lunæ/New and concice Tables of the Motions of the Sun and Moon, by Tobias Mayer, to which is added The Method of Finding Longitude Improved by the Same Author. Commisioners of Longitude, London

Mayer T (1775) Tobiae Mayeri Opera inedita; edidit et observationum appendicem adiecit Georgius Christophorus Lichtenberg, vol I (no further volumes appeared). Joann. Christian Dieterich, Gottingae, see also Forbes (1971c)

Mayer T et al. (1750) Kosmographische Nachrichten und Sammlungen auf das Jahr 1748: zum Wachsthume der Weltbeschreibungswissenschaft von den Mitgliedern der Kosmographischen Gesellschaft zusammengetragen. Krauß, Wienn, Homann, Nürnberg

Meeus J (1998) Astronomical algorithms. Willmann-Bell, Richmond, VA

Meyer W (ed) (1894) Verzeichniss der Handschriften im Preussischen Staate. 1 Hannover. Die Handschriften in Göttingen Bd. 3. A Bath, Berlin

Michaelis JD (1794-1796) Literarischer Briefwechsel von Johann David Michaelis, geordnet und herausg. Von Joh. Gottl. Buhle, vol 1. Leipzig

Moulton FR (1902) An introduction to celestial mechanics. Macmillan, New York; 2nd edn: (1914)

Nauenberg M (1998) Newton's unpublished perturbation method for the lunar motion. Int J Eng Sci 36:1391–1405

Nauenberg M (2000) Newton's Portsmouth perturbation method and its application to lunar motion. In: Dalitz RR, Nauenberg M (eds) The foundations of Newtonian scholarship. World Scientific, Singapore, pp 167–194

Nauenberg M (2001) Newton's perturbation methods for the three-body problem and their application to lunar motion. In: Isaac Newton's natural philosophy, pp 189–224

Newton I (1975) Isaac Newton's theory of the moon's motion (1702): with a bibliographical and historical introduction by I. Bernard Cohen. Dawson, Folkestone, contains facsimile reprint of (a.o.) "A new and most accurate theory of the moon's motion"

Newton I, Cohen IB, Whitman A (1999) The principia: mathematical principles of natural philosophy. A new translation by I. Bernard Cohen and Anne Whitman, with the assistance of Julia Budenz. Preceded by "A guide to Newton's Principia" by I. Bernard Cohen. University of California Press, Berkeley, CA

Pannekoek A (1951) De groei van ons wereldbeeld, een geschiedenis van de sterrekunde. Wereld-bibliotheek, Amsterdam

Pedersen O (1974) A survey of the Almagest. Odense University Press, Odense

Perozzi E, Roy AE, Steves BA, Valsecchi GB (1991) Significant high number commensurabilities in the main lunar problem I: the Saros as a near-periodicity of the moon's orbit. Celest Dyn Dynam Astron 52:241–261

Peters CAF (ed) (1863) Briefwechsel zwischen C. F. Gauss und H. C. Schumacher, vol 6. Esch, Altona

Plackett RL (1958) The principle of the arithmetic mean. Biometrika 45:130–135; Reprinted in E. S. Pearson and M Kendall (eds), Studies in the history of statistics and probability, Charles Griffin & Co., London (1978)

Plana J (1856) Note sur les coefficients théoriques, déterminés par Tobie Mayer, relativement aux deux inégalités Lunaires, ayant pour argument $(2E - 2g + c'm)nt$, $(2E - 2g - c'm)nt$; par Mr. Jean Plana. Astronomische Nachrichten 44:87–90

Press WH et al. (1995) Numerical recipes in C; the art of scientific computing, 2nd edn. Cambridge University Press, Cambridge

RGO (1821) Lunar observations made at the Royal Observatory of Greenwich, from 1783 to 1819. Board of Longitude, London

RGO-NA (1767,...) The Nautical Almanac and Astronomical Ephemeris, for the year .... Published by order of the Commisioners of Longitude, London, various issues consulted

Rice JA (1995) Mathematical statistics and data analysis. Duxbury Press, Belmont, CA.

Rigaud SP (ed) (1832) Miscellaneous works and correspondence of James Bradley. Oxford; Reprinted (1972)

Roy AE (1978) Orbital motion. Adam Hilger, Bristol; 3rd edn: (1988)

Sadler DH (1968) Man is not lost: A record of two hundred years of astronomical navigation with the Nautical Almanac. Her Majesty's Stationary Office, London

Sadler DH (1977) Lunar distances and the Nautical Almanac. Vistas Astron 20: 113–121

Schmeidler F (1995) Astronomy and the theory of errors: from the method of averages to the method of least squares. In: Planetary astronomy from the renaissance to the rise of astrophysics: the eighteenth and nineteenth centuries, Chap 27

Schulze JK (1781) Méthode directe pour déterminer la longitude vraie de la Lune par les mouvements moyens, en se servant de quelques nouvelles Tables qu'on pourroit calculer aisément pour cet usage. Nouveaux Mémoires de l'Académie Royale des Sciences et Belles-Lettres

Seidelmann PK (ed) (1992) Explanatory supplement to the astronomical almanac. University science Books, Mill Valley, CA

Sheynin OB (1971) J. H. Lambert's work on probability. Arch Hist Exact Sci 7: 244–256

Sheynin OB (1973) On the mathematical treatment of astronomical observations (a historical essay). Arch Hist Exact Sci 20:97–126

Sheynin OB (1979) C. F. Gauss and the theory of errors. Arch Hist Exact Sci 20:21–72

Sheynin OB (1993) On the history of the principle of least squares. Arch Hist Exact Sci 46:39–54

Smith GE (1999) Newton and the problem of the moon's motion. In: The principia: mathematical principles of natural philosophy, pp 252–257

Spencer Jones H (1922) General astronomy. Edward Arnold

Stark B (1995) Tables for clearing the lunar distance. Lant-horn Press, Eugene, Oregon; 2nd edn: (1997)

Steenstra P (1771) Grondbeginsels der sterrekunde: of algemeene handleiding om van de allereerste beginselen af tot eene grondige kennis der gantsche sterrekunde te komen. Yntema, Amsterdam

Stigler SM (1986) The history of statistics: the measurement of uncertainty before 1900. Harvard University Press; Cambridge, MA

Stigler SM (1999) Statistics on the table: the history of statistical concepts and methods. Harvard University Press, Cambridge MA

Taylor EGR, Richey M (1962) The geometrical seaman. Hollis & Carter, London

Thoren VE (1974) Kepler's second law in England. Br J Hist Sci 7:243–258; with a supplement by Owen Gingerich and Barbara Welther

Thrower NJW (ed) (1990) Standing on the shoulders of giants: a longer view of Newton and Halley. University of California Press, Berkeley, CA

Tisserand FF (1894) Traité de méchanique céleste, vol 3. Gauthier-Villars

Verbunt F (2002) The earth and moon: from Halley to lunar ranging and shells, unpublished. Available from http://www.astro.uu.nl/verbunt/onderzoek/earth.pdf

Volk O (1983) Eulers Beiträge zur Theorie der Bewegungen der Himmelskörper. In: Leonhard Euler 1707 1783: Beiträge zu Leben und Werk, Birkhäuser, Basel, pp 345–360

Waff CB (1975) Universal gravitation and the motion of the moon's apogee: the establishment and reception of Newton's inverse-square law, 1687–1749. PhD thesis, Johns Hopkins University, Baltimore, MD

Waff CB (1995) Clairaut and the motion of the lunar apse: the inverse-square law undergoes a test. In: Planetary astronomy from the renaissance to the rise of astrophysics: the eighteenth and nineteenth centuries, Chap 16

Walker C (ed) (1996) Astronomy before the telescope. St. Martin's Press, New York

Waters DW (1990) Captain Edmond Halley, F.R.S., Royal Navy, and the practice of navigation. In: Standing on the shoulders of giants: a longer view of Newton and Halley

Whiteside DT (1976) Newton's lunar theory: from high hope to disenchantment. Vistas Astron 19(4):317–328

Wilson CA (1980) Perturbations and solar tables from Lacaille to Delambre: the rapprochement of observation and theory. Arch Hist Exact Sci 22:53–304

Wilson CA (1985) The great inequality of Jupiter and Saturn: from Kepler to Laplace. Arch Hist Exact Sci 33:15–290

Wilson CA (1987) On the origin of Horrocks's lunar theory. J Hist Astron 18(2):77–94

Wilson CA (1989a) Predictive astronomy in the century after Kepler. In: The general history of astronomy, Chap 10

Wilson CA (1989b) The Newtonian achievement in astronomy. In: The general history of astronomy, Chap 13

Wilson CA (1995a) Factoring the lunar problem: geometry, dynamics, and algebra in the lunar theory from Kepler to Clairaut. In: Dumas HS (ed) Hamiltonian dynamical systems: history, theory, and applications, IMA volumes in mathematics and its applications, vol 63. Springer, New York

Wilson CA (1995b) The problem of perturbation analytically treated: Euler, Clairaut, d'Alembert. In: Planetary astronomy from the renaissance to the rise of astrophysics: the eighteenth and nineteenth centuries, Chap 20

Wilson CA (1995c) The work of Lagrange in celestial mechanics. In: Planetary astronomy from the renaissance to the rise of astrophysics: the eighteenth and nineteenth centuries, Chap 21

Wilson CA (2001) Newton on the moon's variation and apsidal motion: the need for a newer "new analysis". In: Isaac Newton's natural philosophy, pp 139–188

Wilson CA (2007) Euler and applications of analytical mathematics to astronomy. In: Bradley RE, Sandifer E (eds) Leonhard Euler: life, work and legacy. Elsevier, Amsterdam, pp 121–145

Wilson CA, Taton R (eds) (1989) Planetary astronomy from the renaissance to the rise of astrophysics: Tycho Brahe to Newton. In: The general history of astronomy, vol 2A. Cambridge University Press, Cambridge

Wilson CA, Taton R (eds) (1995) Planetary astronomy from the renaissance to the rise of astrophysics: The eighteenth and nineteenth centuries. In: The General History of Astronomy, vol 2B. Cambridge University Press, Cambridge

Wolf R (1890–1892) Handbuch der Astronomie, ihrer Geschichte und Litteratur. Schulthess, Zürich, 2 vols; Facsimile reprint Amsterdam (1973)

Young CA (1888) A text-book of General Astronomy for colleges and scientific schools. Ginn and Co, Boston; Revised ed. (1916)

Zinner E (1925) Verzeichnis der astronomischen Handschriften des deutschen Kulturgebietes. Verlag C. H. Beck, München

Index